ENVIRONMENT AND NATURAL RESOURCES MANAGEMENT SERIE
MONITORING AND ASSESSMEN
BIOENERGY
[CLIMATE CHANGE]
ENVIRONMENT

Coping with a changing climate:

considerations for adaptation and mitigation in agriculture

Michael H. Glantz

Consortium for Capacity Building,
University of Colorado and Climate Affairs,
LLC, Boulder, Colorado, USA

René Gommes
Selvaraju Ramasamy

Environment, Climate Change and Bioenergy Division
Food and Agriculture Organization of the United Nations
Rome, Italy

Food and Agriculture Organization of the United Nations Rome, 2009

Reprinted 2010

ISBN 978-92-5-106445-0

FOREWORD

Over one billion people around the world are undernourished because they lack easy and consistent access to affordable food. Climate change is already affecting all four dimensions of food security: food availability, food accessibility, food utilization and food systems stability. The impacts are both short-term, through more extreme weather events, and long-term through changing temperatures and precipitation patterns. Rural communities and livelihoods face immediate risk of increased crop failure, loss of livestock, and reduced availability of marine, aquaculture and forest products and new patterns of pests and diseases outbreak. People living in fragile ecosystems such as coasts, floodplains, mountain areas and semi-arid landscapes are most at risk.

Agriculture, forestry and land use can also contribute to climate change mitigation through reducing greenhouse gas emissions and carbon sequestration. FAO promotes integration of adaptation and mitigation into food security efforts. However, true progress will require comprehensive approaches, close cooperation, synergy and coordination among the policy planners, institutions and local communities.

Adaptation and mitigation strategies should contribute to poverty reduction and at the same time must benefit the most vulnerable communities without harming the environment. Informing about climate change impacts, vulnerability patterns, coping and adaptive capacity as well as facilitating location specific adaptation and mitigation practices are of central concern.

The uncertainties related to climate change impacts and vulnerabilities are often considered as an impediment for concrete and immediate action. However, uncertainty is a fundamental component of climate impacts and cannot, in itself, be used as an excuse for inaction. This document elaborates on issues of less-than-perfect information on climate impacts and vulnerabilities, and need for better informed decisions on "resilient adaptation" by merging adaptation, mitigation and prevention strategies. It offers new perspectives for policy-makers, institutions, societies and individuals on improved ways of identifying most at-risk communities and "best practices" of coping with current climate variability and extreme climate events.

We aim at contributing to approaches and considerations for adaptation and mitigation and improved ways of integrating present-day "best practices" with the longer-term strategies to cope with uncertain future climates.

Peter Holmgren
Director
Environment, Climate Change and
Bioenergy Division,
FAO

ABSTRACT

Changing climatic conditions are projected to affect food security from the local to global level. The predictability in rainy season patterns will be reduced, while the frequency and intensity of severe weather events such as floods, cyclones and hurricanes will increase; other predicted effects will include prolonged drought in some regions; and water shortages; and changes in the location and incidence of pest and disease outbreaks. Growing demand for biofuels from crops can place additional pressure on the natural resource base. New policy driven options are required to address the emerging challenges of attaining improved food security.

The first two chapters of this book presents historical evidence of relationship between climate and food security, as well as current challenges of world food security posed by climate change. The "**introduction**" chapter highlights the need for baseline diagnostics on impacts, vulnerability and resiliency patterns and decision making under uncertainty. **Chapter 2** elaborates on the impacts of climate change on agriculture and stresses how to effectively address these impacts, focusing on ecosystem goods and services and social well being. The chapter on "the setting: baseline information" underlines that mapping, such as capacity to cope in a country, is as important as mapping vulnerabilities to climate variability and change.

Climate change adaptation strategies are now a matter of urgency. Many potential adaptation options in agriculture have mitigation synergies, and similarly, several mitigation options for climate change could generate significant benefits for both food security and adaptation. **Chapter 3** on "Adaptation and mitigation" introduces the "four laws of ecology" and presents their continuing relevance to policy-makers when they identify, develop and implement adaptation and mitigation strategies.

In regard to climate change and the likelihood that future characteristics of climate will change in unknown ways, the existing "best practices" should be viewed as providing a source of tactical short-term response to a changing environment as opposed to untested strategic long-term responses.

Chapter 4 on "What to do at the national level" elaborates the fact that climate impacts and response mechanisms in the near term future are likely to be similar to those of the recent past, barring any abrupt changes in the atmosphere's local and global climatic characteristics.

Most climate impacts of concern to policy-makers are local. Adaptation and mitigation measures, which require poverty reduction and food security, must be customized to benefit the neediest of the needy. **Chapter 5** on "Short-term and long-term policy options" focuses on decision making under uncertainties; improved ways of identifying most at-risk communities and coping with current climate variability and extremes; and improved ways of integrating present-day tactical and "best practice" responses with the longer-term strategic needs.

The conclusion has key **take-home messages** from the FAO high level conference on "World Food Security: The Challenges of Climate Change and Bioenergy" are presented along with closing thoughts about having "no adaptation recommendations without ramifications" as well as suggestions for policy-driven strategic thinking about adaptation to and mitigation of climate change with a focus on improved food security.

Coping with a changing climate:
Considerations for adaptation and mitigation in agriculture

by Michael H. Glantz, René Gommes, Selvaraju Ramasamy

116 pages, 3 figures, 2 tables, 13 pictures
FAO Environment and Natural Resources Service Series, No. 15 – FAO, Rome, 2009

Keywords:
Climate change, bioenergy, food security, adaptation and mitigation in agriculture, coping with climate change in agriculture, short term and long term policy options, policy decisions under uncertainty.

This series replaces the following:
Environment and Energy Series; Remote Sensing Centre Series; Agrometeorology Working Paper

A list of documents published in the above series and other information can be found at:
www.fao.org/nr   and   **www.fao.org/climatechange**

CONTENTS

BANANA CROP DESTROYED BY HURRICANE MITCH (1998) IN HONDURAS

Climate change including extreme events such as storms and floods is making it even more difficult to grow and harvest produce from the land and threatens food security.

EXECUTIVE SUMMARY

- This report is an expanded version of a paper that was originally drafted to encourage participants to the FAO Expert Meeting on Adaptation and Mitigation to provide examples from their regions, sectors and disciplines to reinforce or challenge, as appropriate, the concepts presented in order to improve policy-makers' understandings of and preparations for coping with both the causes and the impacts of climate change on food security.

- The overarching goal in societal responses to climate change for the sake of enhancing food security must be a hybrid strategy, merging adaptation, mitigation and even prevention to produce an overall strategy of "resilient adaptation".

- Governments must decide how they want to systematically think about and then undertake adaptation and mitigation activities. The inherent issues related to national decision making must be evaluated to determine if governments are equipped to cope with the dynamic nature of the impacts of climate change. In other words, are governments able and ready to address twenty-first century climate change problems that are not covered under current policies and programmes?

- Policy-makers are now being pressed to cope with a changing climate, from its anthropogenic causes to its impacts on food security. In this task, they are not unarmed: They can rely on information, knowledge and experience derived from historical accounts of the impacts of climate, water and weather as well as scenarios derived from global and regional modeling activities.

- Many adaptation and mitigation actions to cope with climate change causes and impacts are worth undertaking in their own right.

- Many of the environmental changes that are occurring and those that are likely to occur in the future as a result of climate change are incremental and "slow onset," but they are cumulative. Policy-makers must improve the ways they choose to deal with such creeping changes in the environment as those changes will increasingly influence food security in negative ways.

- A significant number of examples exist of successful and of unsuccessful responses to changes in what we today consider to have been our "normal" climate of the past several decades. Examples of such responses are illustrative of societal vulnerabilities and resiliencies in the face of change, and they serve as a measure of societies' adaptive capacities over time.

- The numerous existing controversies and conflicts in agriculture, forestry and fisheries will most likely be affected by climate change. These controversies and conflicts must be made explicit, and their functional as well as their geographic "boundaries" must be identified and dealt with in a more global and systematic way.

- Adaptation and mitigation activities will be ongoing in order to keep up with changes in the climate, from the global to the local level.

- Adaptation and mitigation activities will generate their own set of impacts on socioeconomic sectors well beyond agriculture, and governments must be prepared to both anticipate and respond to them.

- Policy-makers must beware of short-term, short-sighted solutions. They must also beware of cost-benefit assessments that do not include non-quantitative analyses, such as considerations of social or cultural value conflicts that stem from multipurpose competition.

- Even if policy-makers are on the right track in regard to their development of strategies and tactics for adaptation, they not only have to choose the correct directions in which to move, but they also have to be concerned about the rate of change in the implementation of their policies. American humorist Will Rogers once remarked "Even if you are on the right track, you can still be run over if you are not moving fast enough."

- Prevention strategies and tactics must be pursued along with mitigation and adaptation.

- Do not wait for projections based on the output of scientific models of climate change to confirm what is already suspected or known about the impacts of climate trends, variations and extremes on food security.

ACRONYMS

AAAS	American Association for the Advancement of Science
AFOLU	Agriculture, Forestry and Other Land Use
AOC	Areas of Concern
BAU	Business As Usual
CGIAR	Consultative Group on International Agricultural Research
CI	Conservation International
COP	Conference of Parties
DMUU	Decision Making Under Uncertainty
DRC	Democratic Republic of Congo
FAO	Food and Agriculture Organization of the United Nations
FSIA	Food Security Impact Assessment
GEF	Global Environment Facility
GHG	Greenhouse Gases
GWP	Global Warming Potential
HLC	High Level Conference
HYV	High Yielding Varieties
IFAD	International Fund for Agricultural Development
ILEC	International Lake Environment Committee Foundation
IPCC	Intergovernmental Panel on Climate Change
IRRI	International Rice Research Institute
MA	Millennium Ecosystem Assessment
MDG	Millennium Development Goals
MOD	Manado Ocean Declaration
MSY	Maximum Sustainable Yield
NGO	Non-Governmental Organization
PRECIS	Providing Regional Climates for Impacts Studies
SWOC/T	Strengths, Weaknesses, Opportunities and Constraints/Threats
TECA	Technology for Agriculture
UNDP	United Nations Development Programme
UNFCCC	United Nations Framework Convention on Climate Change
WFP	World Food Programme
WFS	World Food Summit
WMO	World Meteorological Organization
WOC	World Ocean Conference
WOC-CTI	World Ocean Conference and Coral Triangle Initiative
WRI	World Resources Institute

©FAO/X. Van Der Stappen

SMALL-SCALE FARMERS IN AFRICA

Climate change affects everyone. But the worst hit will be hundreds of millions of small-scale farmers, herders, fishers and forest-dependent people who are already vulnerable and food insecure.

INTRODUCTION

CHANGING PERSPECTIVES

The relationship between *climate* and *food security* is obviously not a new issue. In Rome in 1974, for example, the United Nations convened a now-famous World Food Conference under the guidance of the UN FAO. It reminded governments of an urgent need to focus on existing and yet-to-emerge food security and related issues. Thirteen years later (1987), the International Rice Research Institute (IRRI) and the American Association for the Advancement of Science (AAAS) convened an international symposium to address concerns about climate, weather and water impacts on agricultural production. The very same issues of concern to policy-makers today were addressed by scientific researchers then:

The International Symposium on Climate and Food Security ... recognized three critical world problems: that several billion people often lack the most basic human need – food security; that population growth and the need to improve living standards are putting severe pressure on the soil and water resources that sustain all food production; and that unfavorable weather and climate remain the most frequent cause of crop failure – sometimes leading to widespread distress and even famine.

It also recognized a new factor: the growing scientific consensus that the buildup of greenhouse gases in the atmosphere is likely to cause a global climate change – an environmental change on a scale unprecedented in human history – with the potential for great impacts, both beneficial and harmful, on food security.

The overriding concern was: how can scientists help farmers exploit favorable agro-climate patterns and adapt to or protect against unfavorable climatic trends.

In the never-ending struggle to provide people everywhere with the assurance of food security, we certainly need to understand more. But the participants also emphasized the need to apply what we already know – by devising and testing better methods of conveying to farmers the timely and practical agro-climatic information they need.

These were the goals and vision of M. S. Swaminathan, Roger Revelle, and S. K. Sinha – the three eminent scientists who made this meeting possible (Burns, 1989)

In the 1970s no attention was paid to mitigation [at that time mitigation meant the softening of the impacts of an event or process], and concerns about adaptation to climate were centered on weather extremes and climate variability from season to season and year to year to address the crucial aspect of food production stability, one of the pillars of food security. By 1996, however, the World Food Summit (WFS) recognized that the resource base for food, agriculture, fisheries and forestry was under stress and threatened by problems such as desertification, deforestation, over fishing, loss of biodiversity, inefficient use of water, and climate change. Mainly under its commitment three, the WFS made a number of explicit references to the dominant role of climate fluctuations in food supply as one of the main factors interfering with sustainable increases in food production.

Hundreds of meetings and thousands of papers, many of which were focused on climate and the search for food security, have already appeared on societal adaptation to climatic, environmental or societal changes. With such an extensive background, the challenge facing those searching for coping strategies to endure climate change (i.e. global warming) may weigh more heavily on deciding which existing adaptive strategies to pursue rather than on developing yet-to-be-identified unique and untested ones.

In reality, the concept of "food security" has been interpreted in many ways. An FAO report noted that there are more than 200 interpretations of the concept (FAO, 2003; [http://www.fao.org/DOCREP/005/Y4671E/y4671e06.htm]). This report defined food security as follows:

Food security exists when all people, at all times, have physical, social and economic access to sufficient, safe and nutritious food which meets their dietary needs and food preferences for an active and healthy life. Household food security is the application of this concept to the family level, with individuals within the household as the focus of concern (p. 3).

A cursory view of many regions in the world, however, reveals that no matter how one defines the concept of food security, food security as a goal to assure an individual's access to food and nutrition has not yet been realized to any significant extent. This reality has become obvious with the increasing use of and reliance on the term "food insecurity".

Over the decades, the concept of food security has continued to evolve with new twists and turns in its meaning appearing every few years or so. These food security evolutions can be pictured metaphorically as an 'artichoke'. At the heart of the artichoke is the core of the concept of food security, access to adequate nutrition for physical and mental well-being, which always remains the same, but over time different uses of the concept by different users (both individuals and organizations) in pursuit of a wide and varied range of variations on the food security theme to suit their goals and needs add layer upon layer of outer leaves to the center of the artichoke.

Today's concern about climate change has added features to the issue of food security: The acute perception that natural resources are finite (a concept sparked in the late 1960s after the photo was published of planet earth alone in the universe's sea of darkness); that human activities that release greenhouse gases into the atmosphere must be controlled; that adaptation to changing conditions is the most immediate concern for sectors of agricultural production; and that vulnerability to impacts varies greatly from population to population and can even vary in the same location from time to time.

Released in April 2007, the IPCC's 4th Assessment appears to have provided the "tipping point" for governments and many corporations to accept that climate change is a real threat to societies and ecosystems. The global climate has already warmed 0.74 °C since the beginning of the twentieth century. Adaptation concerns are based on the identification of likely impacts of global warming at national, local and household levels and they are increasingly focusing on the development of both proactive and reactive coping mechanisms to soften, if not avoid, those impacts.

The World Bank presents the importance of adaptation in the following way:

Developing countries, and particularly the poorest people in these countries, are the most vulnerable to the adverse impacts of climate variability and ongoing and projected climate change. Their economies depend heavily on climate-sensitive sectors such as agriculture, forestry, fisheries, a reliable water supply, and other natural resources. They are generally hindered by limited human capacity and limited access to technology and capital to invest in risk reduction... Thus it is imperative that climate change adaptation is not separated from other priorities but is integrated into development planning, programs and projects (World Bank, 2008).

Recently, unsustainable development practices for bioenergy production have been recognized as an additional threat and may have an impact on the goal of achieving food security. FAO, in its report on "Food, Energy and Climate: A New Equation" underlined the need to think of food, energy and climate as one interconnected issue.

For millennia agriculture supplied three things: food, fodder and fibre, and played a part in shelter too. Now energy has been added to the list, even if wood has always been used for that purpose. With oil prices near all-time high, governments are supporting the production of biofuels such as ethanol and biodiesal from crops previously grown for food, fodder and shelter. This is helping increase the price of food. (FAO, 2008; [ftp://ftp.fao.org/docrep/fao/011/i0330e/i0330e00.pdf])

THE NEED FOR BASELINE DIAGNOSTICS

Policy-makers need information in order to make the most informed decisions possible. On a weekly basis, however, policy-makers constantly make decisions under uncertainty; that is, they typically do not have the luxury of having in-hand perfect information on which to base their decisions. With regard to the impacts of climate change on agricultural activities, considerable uncertainty remains about the intensity, duration, magnitude and location of impacts, but this uncertainty must not by itself be used as grounds for inaction.

The fact is that climate change-related uncertainties in decision making and Decision Making Under Uncertainty (DMUU) related to food insecurity will likely always exist. This is true because of limitations in our ability to fully understand and therefore predict climate events. Such limitations may become more pronounced as the climate system warms and its behavior becomes increasingly less predictable.

Baseline data are key to an improved understanding of the agricultural impacts of a changing climate and of the rates of change at which those impacts appear. Slow rates of change, for example, provide time for preparation and response, while faster rates provide less time for such actions. Problems will always exist, however, with data, statistics, lack of carbon-adjusted statistics, difficulties in modeling countries' "mitigation potentials," and the still-not-very-well quantified risks of genetic erosion and loss of crop diversity, especially as they occur on-farm. Filling the gaps in baseline data, therefore, is an important aspect of adaptation and mitigation efforts for agriculture and food security.

To facilitate this undertaking, every government needs to undertake a comprehensive, two-pronged assessment of its country's (1) vulnerabilities and (2) resiliencies (defines in this instance as adaptive capacity). Vulnerabilities seem to be relatively easier to identify than are resiliencies. For example, those mired in poverty – children, pregnant women, the infirm and the elderly – are already known to be most vulnerable to hazards and to food insecurity. The same type of assessment is needed for hazard-prone areas such as unstable hillsides, low-lying coastal areas, bushfire-prone areas, and so forth. Resiliencies can be either tangible (e.g. sea walls, effective state of the art early warning systems, available funds) or intangible (e.g. education, training, skills, awareness of risks, perceptive decision making). Assessments such as these can be extremely useful for identifying not-so-obvious vulnerabilities and resiliencies in a society's socioeconomic sectors. As such, there are no targeted activities completed and/or in progress in preparing "resiliency maps" for the vulnerable sectors.

An important aspect of resiliency mapping is traditional knowledge about food production and the nutrition efforts of the world's farmers and herders. Their tactics and strategies had evolved long before recorded history for coping both with variability as well as extremes and even for coping with abrupt as well as incremental change.

A LIVESTOCK HERDER IN TÖV IMAG, CENTRAL MONGOLIA

The natural disasters, known as dzud and drought, affect Mongolia on regular basis causing deaths of millions of heads of livestock and damage significantly the country's economy.

CHAPTER 2

THE SETTING: BASELINE INFORMATION

IMPACTS

The IPCC 4th Assessment

The latest key findings of the IPCC regarding current research results on the state of climate change, its drivers and projections for the future include but are not limited to the following highlights (IPCC, 2007a):

- Warming of the climate system is now unequivocal;
- The rate of warming in the last century is historically high;
- The net effect of human activities since 1750 has been one of warming, due primarily to fossil fuel use, land-use change and agriculture;
- Most of the observed increase in globally averaged temperatures since the mid-twentieth century is very likely (greater than 90 percent) due to the observed increase in anthropogenic greenhouse gas emissions;
- Long-term changes in climate have already been observed, including changes in Arctic temperature and ice, widespread changes in precipitation amounts, ocean salinity, wind patterns and aspects of extreme weather including droughts, heavy precipitation, heat waves and intensity of tropical cyclones;
- From 1900 to 2005, drying has been observed in the Sahel, the Mediterranean, southern Africa and parts of southern Asia;
- More intense and longer droughts have been observed over wider areas since the 1970s, particularly in the tropics and subtropics;
- Continued greenhouse gas emissions at or above current rates would cause further warming and induce many changes in the global climate system during the twenty-first century that will very likely be larger than those changes that were observed in the twentieth century;
- Projections for the twenty-first century include a greater chance that more areas will be affected by drought, that intense tropical cyclone activity will increase, that the incidence of extreme high sea levels will increase, and that heat waves and heavy precipitation events will be more frequent; and

■ Even if greenhouse gas concentrations were to be stabilized, anthropogenic warming and sea-level rise would continue for centuries due to the timescales associated with climate processes and feedbacks.

The IPCC 4th Assessment and food security

This IPCC Assessment addresses food security by discussing the foreseeable impacts on agricultural productivity and production in different regions around the globe. The report's collective comments suggest that some areas will benefit from global warming, at least through a transitional period, though most areas will be adversely affected. Significantly, the assessment emphasizes that those areas that do benefit from global warming in the near to mid-term will eventually also suffer from declining productivity. Various parts of the assessment also reference changes in the hydrological cycle that will affect agriculture in general and food security specifically. Migrations forced by climate change (for example, excessive heat, increased evaporation rates, or prolonged drought-induced crop failures, or flood) will further burden the already stretched agricultural resources and food supplies of regions that have managed to sustain productivity.

While each region around the globe will have to develop its own adaptation, mitigation, prevention and response strategies, inhabitants of the African continent will likely be the most affected and most needful of resources, if they are to effectively respond to climate change:

> *Agricultural production, including access to food, in many African countries and regions is projected to be severely compromised by climate variability and change. The area suitable for agriculture, the length of the growing seasons and yield potential, particularly along the margins of semi-arid and arid areas, are expected to decrease. This would further adversely affect food security and exacerbate malnutrition in the continent. In some countries, yields from rain-fed agriculture could be reduced by up to 50 percent by 2020. (http://timeforchange.org)*

The IPCC's 4th Assessment is the culmination of a process that began over twenty years ago in the late 1980s. Preceded by the IPCC's 1st, 2nd and 3rd Assessments, the trends in greenhouse gas emissions and global warming's likely impacts as noted in the 4th report of the IPCC are consistent with trends that were reported in those earlier IPCC assessments, with each new assessment having further bolstered the evidence for human contributions to the naturally occurring greenhouse effect. Making a bad situation appear even worse is the evidence that the rates of several environmental changes, such as the melting of Arctic sea ice, have actually accelerated in recent years.

A climate change challenge for society: riding the variability curve

The 4th Assessment clearly played a key role in the sharp, step-like increase in concern over the climate change issue after its release in 2007, in many ways proving to be the "tipping point" for policy-makers worldwide who truly began to take more seriously the climate situation after its release. Recognition of the IPCC process with the awarding of the Nobel Peace Prize served to enhance the influence of the 4th Assessment, especially with the broader public. Concern over climate change has sparked an unprecedented "rush to action." Though deserving of such focus and concern, governments and other climate-, water- and weather-related scientific research and application funding agencies must beware, especially with regard to their response to this one climate change report, of the likelihood of "overshoot"; that is, agencies must beware of over-focusing on what has become the most popular and recognizable concept in climate reporting, "change," and risk neglecting other important, less reported climate factors – such as variability from season to season, year to year, and decade to decade – that have often not been record setting anomalies but have none-the-less had serious consequences for societies and ecosystems. After all, the broad notion of climate change includes variability in the set of such climate factors as temperature, which will change at different rates; changes in the expected flow of the seasons; and changes in the timings, intensities and locations of precipitation.

Concern about the potential occurrence of an abrupt climate change tends to draw attention away from possibly substantial transformations in the naturally occurring variability of our existing, relatively well-understood global climate regime. Societies, their institutions and the individuals that compose them have always struggled to understand and forecast variability on various time scales, especially the seasonal and inter-annual ones, either to take advantage of good climate conditions or to prepare for adverse ones. This can be referred to as an attempt by societies to "ride the climate variability curve." In any of the climate change scenarios set forth thus far, variability will continue; however, given that the future state of climate is uncertain, such variability may shift beyond the bounds of an anticipated range, resulting in unexpected climate scenarios. Precaution should be taken to compensate for possible upturns and downturns in climate variability in order to be better positioned to prevent or mitigate the impacts of these unknowns. The fisheries sector provides perhaps one of the most straightforward examples of this response to variability.

Fish populations vary from year to year, with some species exhibiting high variability in reproduction because of environmental factors combined with recruitment processes. A perfect fishery would, arguably, enable the fishing community to ride the seasonal variability curve's ups and downs; however, forecasts are not good enough to allow for such a perfect scenario, so fishing strategies must include a range of management options such as Maximum Sustainable Yield (MSY), optimal yield and safe yield. Maximum sustainable yield is an attempt to eke out the maximum level of fish catches possible. For this management strategy, however, the risk of over-fishing or of a collapse of the fish population is high due to fish population dynamics and populations' interactions with environmental variability. Optimal yields can be viewed as a compromise to split the difference between the risk-averse safe yield approach and the risk-taking MSY approach. Safe yields have the lowest probability of fishing pressures destroying fish populations, but it also provides the lowest level of potential catches. The management strategy for fisheries for a given place must reflect a level of caution (e.g. level of fishing effort), given the numerous uncertainties that can surround the exploitation of living marine resources.

A perfect forecast of variability a season or two in advance would allow farmers and other stakeholders to prepare well in advance for shifts in climate conditions. Such preparations might include lowering stocking rates on rangelands if drought is forecast; more or less stringent controls on fishing limits; planting shorter season grain varieties or completely shifting to better suited crops, and so forth. Without such perfect forecasts, however, skills in the form of education and training combined with existing "ordinary" knowledge become necessary for effective management of climate-sensitive resources related to food security. Regardless, societies must not shortsightedly focus only on 'change' that might occur in an unspecified distant future, but must continuously improve their ability to cope with seasonal and inter-annual variability as well as decade-scale fluctuations as the climate warms, altering the climate variability that we have become accustomed to in our experiences of the recent past.

Does climate impacts history have a future?

Most people tend to value present-day events and knowledge more highly than past events and knowledge and possible futures. Economists call this discounting; one euro in the pocket now is worth more than the same euro in the same pocket several years from now, according to this economic principle, because, put simply, people have to survive the present in order to participate in the future. The problem with this standard for valuation is that a considerable amount of usable knowledge exists in the records and folk wisdom of people from the generations that preceded ours. Learning about how climate, water or weather anomalies affected food security in the past and how societies coped or failed to cope can provide usable insights into how to respond to similar or analogous impacts in the future.

The problem is that many people (researchers as well as policy-makers) tend to believe that such historical information has become outdated because of scientific, engineering, or technological progress and because lessons about coping with disasters were learned. As a result, historical climate-, water-, and weather-related impact information, even information about recent impacts, is often neglected, even though such information could often provide context and guidance for present and future planning. The impacts of anomalies on food security in the recent past, for example, will possibly produce similar impacts in the near term. While speculating about future impacts, therefore, these historical accounts must be exploited in developing adaptation strategies to cope with these issues at local to national levels.

ASPECTS OF VULNERABILITY
Ecosystem changes

Considerable attention has focused on the IPCC assessment process, which began in the late 1980s. What has been as important in a different way has been the recent release of the Millennium Ecosystem Assessment (MA). The website for the MA fully explains its origin and importance (MA, 2005; [http://www.millenniumassessment.org/en/About.aspx]), though an excerpt here is useful:

The Millennium Ecosystem Assessment (MA) was called for by the United Nations Secretary-General Kofi Annan in 2000. Initiated in 2001, the objective of the MA was to assess the consequences of ecosystem change for human well-being and the scientific basis for action needed to enhance the conservation and sustainable use of those systems and their contribution to human welfare. The MA has involved the work of more than 1 360 experts worldwide. Their findings ... provide a state-of-the-art scientific appraisal of the condition and trends in the world's ecosystems and the services they provide (such as clean water, food, forest products, flood control, and natural resources) and the options to restore, conserve or enhance the sustainable use of ecosystems.

Given the central importance of ecosystems to societal well being, some key observations about the risks associated with "tampering" with the sustainable functioning of ecosystems are instructive.

About 25 years ago a schematic diagram, reproduced in Figure 1, presented an idealized picture of a food production system.

The figure suggests that weather affects only crop yields; however, even at that time weather's effect on many of the boxes in the graphic was well known. Weather's broader influence is suggested in another version of the graph (Figure 2), in which the box previously marked as "weather" is replaced by "drought."

In fact, lines in Figure 2 can be drawn from the drought box to many of the boxes in the diagram – even the "tastes" box – as humanitarian food imports of wheat or yellow corn, not being the staple of the food importing region, have even been known to distort local food preferences. This situation has led to arable land being removed from traditional crop cultivation and given to cultivation of non-traditional, climate-sensitive food crops.

In addition to what is already known or what will likely be the impact of episodes of extreme weather and climate on food production and, therefore, on food security, it is reasonable to speculate on the major impacts that might accompany global warming. In truth, such speculation has already been happening for several decades. The most legitimate assumption is that every box in the above graphic would be affected if the weather box were replaced by a "global warming" box.

Beyond serving as interesting illustrations of the point, these diagrams also underscore what has been called the Four Laws of Ecology and the basic belief that in nature "you can't change just one thing." Taking this law into account,

FIGURE 1

Schematic diagram idealizing a food production system (Glantz, 1987; originally published by the US Department of Agriculture in 1984 -"sub-Saharan Africa: outlook and situation report, Economic Research Service).

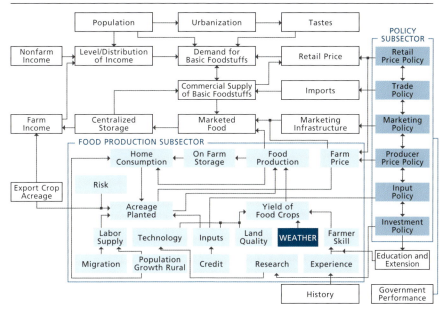

FIGURE 2

Schematic diagram in which 'drought' replaces 'weather' as the affecting parameter

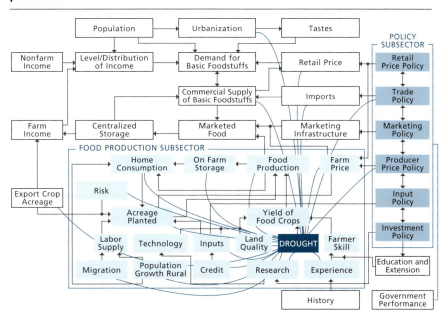

a key issue for governments is that of ensuring that billions of people around the globe with little purchasing power have access to and receive adequate nutrition (i.e. food security at the household level), while preserving the planet's biodiversity, which is at the root of the sustainability of life on earth. While the interactions and values involved in this issue are complex, complexity, like uncertainty, cannot be used as an excuse for inaction or used to exclude elements of civil society from participating in planning for the climate change future.

One way societies can attend to the complexities of these ecosystemic issues is by paying attention to the value (and pervasiveness) of "usable" ordinary knowledge." Lindblom (1979) referred to "ordinary knowledge" as:

> *knowledge that does not owe its origin, testing, degree of verification, and truth status to current distinctive [research] techniques but rather to common sense, casual empiricism, or thoughtful speculation and analysis. It is highly fallible, but we shall call it knowledge, even if it is false. As in the case of scientific knowledge whether it is true or false, knowledge is knowledge to anyone who takes it as a basis for some commitment or action...For social problem solving, we suggest people will always depend heavily on "ordinary knowledge."*

As the saying goes, "knowledge is power. Sharing knowledge is empowering"; the task of researchers and policy providers, therefore, is to assure the correctness of the knowledge base that is passed on to individuals in society. Their task is also to become empowered by learning from local knowledge that had been garnered through trail and error over long periods of time.

The conclusion of the Millennium Assessment about societal well-being and ecosystems goods and services suggests that in order for ecosystems to have value or merit protection from destruction they must provide tangible goods and services to society. A provocative, new understanding emerges, however, when the two ideas central to the MA conclusion are rearranged to read as follows: in order for human goods and services to have value or merit protection they must provide tangible benefits for ecosystems' well-being. In other words, human activities must be pursued with the sustained well-being of ecosystems as a key objective. Although composed of the same two ideas, these converse notions for the new millennium and a changing climate would yield very different outcomes for both societies and ecosystems.

BOX 1

A PRECAUTIONARY NOTE ON DEFINITIONS

Discussions about climate variability, climate change, climate extremes and the impacts of each on societies and ecosystems are filled with such terms as coping, capacity of response, vulnerability, resilience, adaptive capacity, sensitivity, adaptation, mitigation and - rarely these days – prevention. An important (troublesome, actually) problem with the concepts typically used in climate change discourse was, however, analyzed by Latin American researcher Gallopin (2006). He noted the following:

The terms vulnerability, resilience, and adaptive capacity are relevant in the biophysical realm as well as in the social realm. In addition to being terms in colloquial language, they are widely used by the life sciences and social sciences, not only with different foci but often with different meanings...Sometimes the concepts are used interchangeably or as polar opposites... This plurality of definitions is possibly functional to the needs of the different disciplinary fields... but sometimes it may also become a hindrance to the understanding and communication across disciplines.

Gallopin (2006) went on to "attempt to highlight the fundamental attributes of the three concepts and to identify the conceptual linkages between them." Still, the reality is that popular usage of these terms and other synonyms will rule the day, regardless of how hard academic researchers seek to clarify their meaning [NB: it is important to note that the UNFCCC and the IPCC do not use the same definition of such a central concept as "adaptation" [(Pielke, 2003; www.climateadaptation.net/docs/papers/pielke.pdf)]. This is the situation with which researchers and decision makers will have to live and, more importantly, of which they must continuously be aware.

©FAO/B. Giorgi

A FARMER IN THE MOUNTAINOUS AREA OF THE VALLEY OF GUILIN IN GUANGXI, CHINA

Mountains are early indicators of climate change. Extreme events are likely to become more common and more intense in mountain areas, threatening the livelihoods of both mountain people and those who depend on mountain areas for water, food and other resources.

Vulnerability patterns

Global statistics, like statistical averages, are useful for a wide range of purposes. For example, researchers talk in terms of the global average temperature having increased by 0.74 °C since the beginning of the 1900s. This is doubtless a useful piece of information to alert people that the global atmosphere is on a warming trajectory; however, it only represents a *global* average of regionally warmer and cooler locations worldwide. Yet, national policy-makers need regional and local information in order to make policy decisions relevant to their citizens and their country's climate-related hazards. The same problem exists with demographic statistics. Global averages and global rankings using vulnerability indices for food security, for example, are useful for some purposes but may not be useful for national policy-making purposes associated with climate change-related adaptation, mitigation and prevention. National policy-making, on the other hand, requires country-specific information, such as demographics as who and which regions are most at risk to climate variability and extremes, much of which is already available but may not be readily accessible or in a centralized location.

As argued elsewhere, who is vulnerable to climate variability and extreme climate, water and weather events is generally known, and this knowledge can be directly correlated to the most likely victims of climate change. However, a breakdown provided by socio-economic and livelihoods groups, by geographic area, by farming systems or by sub-sectors will further help policy-makers to identify at-risk groups. Of special relevance is the state of the world's crop diversity, as it plays a major part in adaptation to climate change for livelihood measures.

Vulnerability is generally defined as a function of risk and exposure. Vulnerability with regard to climate change implies that people are exposed to aspects of climate that are changing in ways that will either generate or increase risk, which generally implies a potential loss of something valued. For food security, the risk is of poorer nutrition or reduced access to food supplies than would be expected under "normal" climate conditions. The capacity to cope with the risky situations under a given exposure to hazards (both natural and human induced) also shapes the pattern of vulnerability. As often is the case this capacity is weak in the part of the world that suffer from food insecurity either intermittently or chronically.

Resiliency patterns

Resilience, which has several definitions but generally refers to the ability of a society to "bounce back" after suffering an adverse impact, is sometimes viewed as the opposite of vulnerability, but it really isn't. The impression that these are opposing terms derives from the mistaken idea that resilience entails a fundamental robustness, whereas vulnerability suggests fragility. However, is the ability to 'bounce back' to a condition that was unsustainable or unsound to begin with really the resiliency societies or groups should strive for after an adverse impact? Does such a situation really demonstrate a fundamental robustness? Or is true robustness of a people represented by their ability to 'bounce back' from adversity to an improved condition over the one that had previously existed?

Resiliency viewed as the ability to "spring back" from and successfully adapts to adversity, is also used to indicate a characteristic of resistance to future negative events as commonly referred in human stress related psychology and strategies at personal, organizational and leadership levels in business and management field. The IPCC (2007a) defines "resilience" as the ability of a social or ecological system to absorb disturbances while retaining the same basic structure and ways of functioning, the capacity for self-organization, and the capacity to adapt to stress and change. Resiliency can also be defined by a capacity to cope successfully in the face of significant future risk. Mapping such a capacity to cope in a country is as important as mapping vulnerabilities to climate variability, extremes and change because such baseline data facilitates an understanding among planners and policy-makers of where risk is most critical.

As described on a management oriented website (AlphaThink Consulting, 2003), resiliency maps are already undertaken for individuals.

Essi Systems' Resiliency Map will help you explore your resiliency demands, assets and current levels of functioning. The Resiliency Map pinpoints your strengths and vulnerabilities, detects areas of caution and strain, and helps you chart new strategies for enhancing personal health and overall performance.
[http://alphathink.com/Frame-944278-servicespage944278.html?refresh= 1193338038490]

Such narrow-scale mapping could be used to evaluate a household, village, region or country's ability to recover to near "normal" or improved food security conditions following an adverse impact.

Rates and processes of change

Regarding adaptation to global warming's impacts on agriculture, fisheries, forestry, health, public safety, and food security, some of the most important factors are the expected changes to the rates at which various key aspects of climate change – rainfall, temperature, relative humidity, cloudiness – and at which evapotranspiration, the process by which moisture is exchanged between the atmosphere and vegetation and soils, occurs. If the rates change incrementally and societies are aware of those changes, those societies may be able to adjust human activities accordingly. Within limits, some ecosystems will likely also be able to adjust to incremental changes. If, however, the rates of change are too rapid to be viable for adjustments like shifting agricultural practices, changing crop rotations, developing new fodder regimes for livestock as grasslands dry out, then societies will be unable to escape with minimal impacts to their climate-sensitive activities and to the ecosystems on which those activities depend.

Virtual water and ghost acres

All reports on the hydrologic cycle suggest that the cycle will intensify as the atmosphere warms, with some suggesting that the cycle could yield about 15 percent more precipitation per annum. At this point, however, conjectures based on global circulation model output are little more than speculation and educated guessing, not yet reliable enough to predict with any accuracy where the precipitation would fall, how it might fall, or when it will fall. Paradoxically, these reports also suggest that water scarcity in the next couple of decades is highly probable, with extreme shortages already appearing in various locations around the globe. As changes to the global water cycle become more pressing, policy-makers will have to scrutinize more closely where their limited water supplies are going and what they are being used for. The concepts of virtual water will become more and more relevant as these cycles continue to change.

Virtual water is calculated in terms of the water that is used to grow crops that are exported to (or imported by) other countries. According to the concept, water used to grow flowers in Kenya, for example, is actually calculated as supplemental water supplies of the countries that import those flowers. In this manner, Kenya's water resources are not being used for its domestic food and energy needs. As another example, a country that imports wheat instead of producing it on its own soil is, in essence, borrowing water supplies from another country's water supply that had been used to produce the wheat. Governments around the world must reevaluate both their water and food balance demands and supplies in terms of 'virtual water'. Understanding the notion of 'virtual water' can enable a government to better understand where its finite water resources are being consumed and for what purposes.

Similarly, the concept of ghost acres (or ghost hectares) was developed several decades ago. It was used to explain that food imports by Country a relied for those imports on the agricultural lands of Country B. In the same way, the "Green Revolution" also provided ghost acres in that the use of fertilizers and irrigation enhanced agricultural productivity and overall production from beyond what the land might have been able to provide in its natural state (Lang and Heasman, 2004). The notion of ghost acres has also been applied to protein taken from the sea, which serves to supplement the protein produced on the land. A country such as Japan, for example, would require several times more farmland than it has in order to produce an equivalent amount of protein to replace the amount it takes from the sea. The notion of ghost acres also applies to a country's food imports as well.

Global warming and disappearing seasons (as we've come to expect them)

The disappearance or even the change in the overall characteristics of a season (i.e., seasonality) should concern everyone. What else might change, related to changes in the seasonality to which people have become accustomed to in their regions? For example, over the past decade, the ice on various lakes in the northern central United States was no longer strong enough to support ice fishermen and their equipment.

For years, the expected patterns of the seasons have been shifting almost imperceptibly. Those seemingly small changes have, over time, however, accumulated to become more and more visible, leading to seasonal flows in different locations around the globe that societies have been accustomed to. Winters have, in general, become drier and warmer in many regions, and spring rains now come less predictably, both in timing, in frequency and in intensity. Multiyear droughts in Australia and the southeastern United States have generated concern about the "aridification" or the drying out of these regions.

The disappearance or even substantial changes in the overall characteristics of the four seasons as they are expected should concern everyone. The problem is that over the past few decades, winters have in general become drier and warmer in many regions. Rainy seasons have become less so, not abruptly but incrementally over time. Both industrialized as well as developing economies and economies in transition live by the expected flow of the seasons, so no country will escape changes in seasonality with a warming atmosphere. Such changes will affect human settlements worldwide in ways that most communities are just beginning to consider. For example, researchers predict chronic water shortages worldwide (as in the Eastern Congo), a shifting boundary between rangeland and farmland, recurrent and prolonged drought (as in various parts of sub-Saharan Africa, Australia and Southeast US), a potential increase in the number and frequency of famines and perhaps a shift in their locations, and a shortening or lengthening of local and regional hazards related to climate, water, and weather. Adaptation strategies need to focus on this high priority aspect of climate change.

An aspect of the consequences in terms of food security, specifically, of the impacts of global warming includes but is not limited to the following: changes in the growing seasons' length as well as the timing and amount of precipitation; changes in the snowfall season, the runoff season, the rainy season, the timing of flood recession farming, the hunting season, the fishing season, the water season, changes in the timing of outbreaks and increases in vector-borne diseases, rice farming following the replacement of saline water intrusion in rivers by freshwater after onset of rains (e.g. Mekong River), extended seasonal food crisis because of long-lasting drought conditions (e.g. "*Monga*" in Bangladesh), and so forth. Speculation about the foreseeable impacts of changes in seasonality is virtually boundless.

©FAO/R. Grisolla

CHILDREN IN THE MOUNTAINS OF PERU

The world's population is young, with nearly 2.2 billion people under the age of 18. Children and young people have enthusiasm, imagination and abundant energy to undertake local actions to manage climate risks.

APPROACHES TO IMPACT ASSESSMENTS
Forecasting by analogy:
The future is here for those who wish to see it

Many of the adverse climate-change-related environmental scenarios people have discussed, especially regarding the consequences of future human interactions with various types of ecosystems, from deserts (i.e. desertification) to mountain slopes (i.e. deforestation), have already been occurring for decades. Such scenarios should, therefore, no longer be viewed as speculation because the impacts of those changes have already been demonstrated, if not within one country, then within another. Even where there is a paucity of data for one particular area, the results of similar modifications to the natural environment have already been tracked and tested in other areas, yielding results that have demonstrated these modifications as being either good or bad for the environment, for society, or for both. Such correlations are at the heart of "forecasting by analogy."

The deforestation of mountain slopes, for example, will likely yield results in remaining forested mountain areas that are similar to those that have been witnessed in areas where such degradation has already taken place; in other words, the experiment of mountain slope deforestation has already been performed and the results are in hand, at least as far as the long-term impacts on the natural environment are concerned. When similar approaches to mountain forest management are attempted anew in a similar topographical setting elsewhere on the globe, therefore, similar results – soil erosion, rapid rainfall runoff, lower soil moisture recharge, sediment loading of streams, dams and reservoirs, and faster snowmelt in the spring – should be expected.

Prolonged dry spells and especially severe droughts expose inappropriate land use practices of farmers and herders; that is, practices that are inappropriate during periods of moisture stress but that are hidden or tolerated by nature during periods of favorable rainfall. A similar situation is likely to occur with regard to climate change, as the various characteristics of climate intensify or shift to locations where they had not before been witnessed. Policy-makers and individuals alike need to be alert to subtle changes in the environment or in the human interface with climate-sensitive ecosystems. It is also important to be aware that severe droughts can expose sustainable land management practices. The process of forecasting by analogy is valid when considering scenarios for other ecosystems, like the destruction of mangrove forests for the development of shrimp ponds or the irrigation of soils in arid areas without putting proper drainage facilities in place.

While some governments have made sustainable changes to their environments, others have not. The point is that "new" scientific assessments of potential environmental impacts for each and every human interaction with the environment are often not necessary because the impacts of most human-induced environmental changes have already been sufficiently demonstrated. The truth is that calls for new environmental impact assessments are sometimes used as delaying tactics by those who will benefit, often for corporate, political, or personal gain, from their proposed changes to the environment. The bottom line is that the future environmental impacts of some of these new activities already exist somewhere on the globe, if only we would choose to see those inevitable futures and take proactive action accordingly.

Making hotspots visible

"Hotspots" has become a popular term in recent years, increasingly being used to draw attention to particularly calamitous situations in ecosystems around the world. The term also has social contexts, by positively highlighting such concepts as cultural hotspots, skiing hotspots, tourist hotspots, scuba hotspots, and so forth. "Hotspots" is, in this manner, a somewhat awkward notion, because it evokes a location or an activity or a situation that is beyond the usual or out of the ordinary, but whose specific meaning, which is accentuated as either very positive or very negative as a function of the term itself, is wholly dependent on its context in any usage.

For our purposes, "Hotspots" can be defined as locations or activities of interest to a group or organization where human interactions with the environment are considered to be adverse to the sustainability of an ecosystem or those human activities that are dependent upon it. It is a segment along a continuum of environmental change. For FAO in particular, "Hotspots" refers to adverse aspects of the interface between agricultural activities and environmental processes. These definitions are purposely rather broad to enable points of entry into an FAO-wide hotspots programme for activities related to agriculture, forests, fisheries, food security and nutrition (Glantz, 2003 [ftp://ftp.fao.org/docrep/fao/006/y5086E/y5086E00.pdf]).

Some might consider the recent surge in the use of the term "hotspots" to be a new environment-related fad; regardless of how it is viewed, however, the concept can be used to identify situations that, if left unattended, could prove harmful, both to the environment and to those dependent on it. The reality is that every country needs to prioritize its hazards in terms of the likelihood of their occurrence and the severity of their impacts on the people, infrastructure and ecosystems. No country, rich or poor, industrial or agrarian, capitalist or socialist, can address at once all of its "areas at risk" of hazards by putting into place adaptation mechanisms to protect them, so adaptation measures in most cases have to be implemented "in parts," with the highest priorities given to the protection of those areas of greatest concern to both the government and civil society.

One of the truly global "hotspot" aspects of climate change is sea level rise. All island nations as well as low-lying coastal areas are at high risk of suffering from this aspect of climate change. Unlike shifts in rainfall or changes in seasonal characteristics, a rising sea level yields only losers. And the options available to individuals and governments, on local to national scales, to adapt to this aspect of climate change are few and costly – retreat from the low-lying coastal areas, re-enforce coastal barriers to the sea and its surges, voluntary or forced abandonment of the at-risk area.

Another foreseeable hotspot to expect in mid-latitude regions around the globe is the emergence of tropical vector-borne diseases. Mosquitoes, for example, do not respect political borders and can easily spread pole-ward away from the equator into regions where the parasites they carry had not been present before. Infectious diseases such as malaria and dengue fever have always been seen as tropical or developing country problems, but they will increasingly become a concern to industrialized countries in the mid-latitudes as the temperature of the atmosphere rises.

Conservation International (CI) has produced an interactive map that identifies biodiversity (biological and soil) hotspots – in this context the term is wholly negative, referring to locations at risk to biodiversity loss – around the world by continent. Interestingly, the notion of "bright spots" was introduced on the same map, identified as areas where the degradation of soils has either been arrested or reversed.

[http://web.biodiversityhotspots.org/xp/Hotspots/hotspots_by_region/]

FIGURE 3

Hotspots Pyramid showing an idealized progression of environmental change

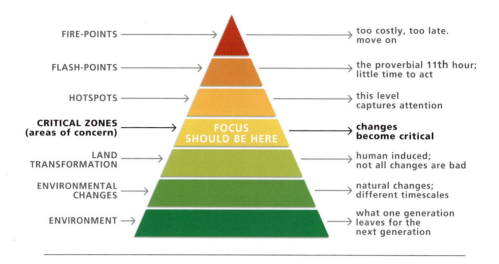

FIRE-POINTS → → too costly, too late. move on

FLASH-POINTS → → the proverbial 11th hour; little time to act

HOTSPOTS → → this level captures attention

CRITICAL ZONES (areas of concern) → **FOCUS SHOULD BE HERE** → **changes become critical**

LAND TRANSFORMATION → → human induced; not all changes are bad

ENVIRONMENTAL CHANGES → → natural changes; different timescales

ENVIRONMENT → → what one generation leaves for the next generation

The hotspots pyramid and adaptation
Areas of Concern (AOC)

Figure 3 can be referred to as a hotspots pyramid. A glimpse of the pyramid shows, in graphic form, a simplified progression of environmental changes (forests, irrigated lands, rainfed cultivated areas, fisheries, etc.) that can result from environmental interactions with human activities. Assume, for example, a swath of land in a pristine state. As humans move in, they begin to transform the land. Cultivators prepare it for food production, and herders graze their livestock on its grasses. Notably, not every such interaction between agriculture/herding and the environment is a negative, non-zero sum one, where either agriculture wins or the environment wins and the other loses. If developed sustainably, both agriculture and the environment can prevail in a sustainable way.

As unsustainable transformations to the land accumulate and the land becomes exhausted from overuse, however, cultivators and herders increasingly move into marginal areas with poorer soils and more erratic rainfall in an attempt to maintain or even increase production. Changes in the land become much more pronounced but are not seen to be of crisis proportion by policy-makers who are busy dealing with other, more urgent socio-economic problems, especially since, at least in the short term, the cultivators and the herders are able to maintain production levels by encroaching further and further onto more and more marginal tracts. At this point, this once pristine swath enters a critical zone. The changes that have accumulated have brought the ecosystem to the brink of collapse.

Hotspots are locations of degradation of either the managed or the unmanaged environment. They indicate situations when mitigative action remains possible at relatively higher costs and appear before conditions deteriorate further to the flashpoint stage, the proverbial 11th hour when actions to restore environmental quality before long-term, irreversible destruction occurs. The Flashpoint stage offers a brief, last window of opportunity for policy-makers to react before environmental collapse becomes inevitable; when (if) they do choose to react, however, the necessary measures for recovery will prove extremely costly in terms of time, money and political capital. Firepoints indicate that environmental conditions have collapsed – it's too late for policy making, as the degradation has overwhelmed all chances for recovery, and exhausted fields, for example, have to be abandoned for generations to come.

Typically, only when a swath of land (or section of the sea) has been labeled a Hotspot, indicating that a crisis situation has emerged, does it begin to receive the serious attention of local officials and the national media. Such reactions are decidedly ill-timed, however, as more proactive attendance to foreseeable and developing crises would prove beneficial to all stakeholders, and especially to the environment itself. The fact is that policy-makers should focus not on Hotspots but on Critical Zones (Areas of Concern) because at this stage of the continuum (represented graphically in the Hotspots Pyramid) not only does enough scientific evidence of degradation exist but so does enough lead time to proactively implement relatively low cost yet highly effective measures to arrest or reverse the devastation and avoid the negative consequences (again, both socially and environmentally) that accompany passage into the Hotspots stage. Indeed, Areas of Concern merit considerable attention as indicators of adverse change and as a focus for policy discussion and execution.

Policy-makers do not have to wait for Hotspots to appear before they take preventative action they can respond when early warning observers, who are key participants in a comprehensive early warning system, alert them of an emerging Areas of Concern. The primary objective of a focus on food-security-related hotspots is to avoid creating them where they do not yet exist. Most such changes to the environment by human activities are of the slow onset but accumulating kind. The hotspots pyramid can be used to discuss changes in agricultural activities and in the tendency to slide from a state of food security toward one of food insecurity.

Creeping environmental change

Quick-onset changes in climate and the environment are easy to see but difficult to cope with. Slow-onset changes, on the other hand, are difficult to see and even more difficult to cope with, at least in a timely way. Crop failure due to drought occurs over a short time and is obvious to the observer. Decline in crop yield, however, is more readily detected over a longer time period. Governments in general tend to have considerable difficulty dealing with slow-onset, low grade but cumulative changes to the environment. The same holds true for similar creeping changes in both managed and unmanaged ecosystems as well as for changes in various aspects of climate, including subtle changes in temperature, rainfall, inter-annual variability, record-setting anomalies and so forth. Governments need to spend more attention coping with creeping changes in climate, water and weather because those incremental creeping changes eventually accumulate, leading to crises at some time in the future. For example, "famine" can be viewed as either an event or a process. Perceived as an event, famine is usually identified, on the one hand, in terms of the number of people forced to seek food in refugee camps. As a process, on the other hand, famine is identified by indicators of progress (change) that constitute subtle indicators along the path toward famine, such as increased sales of personal property (e.g. jewelry or cooking pots), the drastic forced thinning of herds and unfavorable market behavior of land, livestock, credit and water each of which works against the scarce resources of poor farmers and herders.

The 4th Law of Ecology states that there is no "free lunch" (see section 3.1), and this law holds true when it comes to neglecting creeping environmental changes, regardless of cause, whether natural or human-induced. Creeping changes, by their very nature, accumulate and eventually become major changes, which usually materialize in environmental crises that interact with – if not create – other creeping environmental changes.

For example, deforestation of mountain slopes can lead to soil erosion and increased runoff during heavy rains, intensifying the turbidity loads of rivers and streams. This silt continues to build up until it settles in reservoirs and behind dams, decreasing their utility and shortening their expected lifespan. This situation, in turn, reduces the amount of water that the dam or reservoir can provide to downstream users, while the increased runoff can lead to more serious and more frequent flooding of settlements and cultivated areas.

Global warming as a creeping environmental change

Climate has been changing slowly since the early 1900s. This is an accepted scientific fact; by the year 2000, the global average temperature had increased by 0.74°C. For three-quarters of the twentieth century, these changes were noted with little fanfare, and scientists could not even determine with a degree of certainty why the climate was changing. In the mid-1970s, however, convincing evidence that is now considered reliable for the likely human-induced cause of changes in global climate – the increasing levels of greenhouse gas emissions into the atmosphere – began to accumulate.

In retrospect, those responsible for food security in their countries and at the global level have for some time now been engaged in food-security-related decision making under uncertainty. For the most part, they did so in the distant to near past by responding to seasonal and inter-annual forecasts. More recently, however, their decisions have been made under a different set of conditions: They now know that certain human activities are in part responsible for enhancing the naturally occurring greenhouse effect; they now know that global temperatures are reaching levels not seen in tens of thousands of years; they now know that global warming will likely bring impacts that have not been witnessed in human history; and they now know that these physical changes are taking place at a time when nearly 7 billion people are dependent on the earth's limited resources for their lives and livelihoods.

The climate has been changing slowly for some time now (creeping along incrementally but cumulatively), and policy-makers and especially local farmers and herders have been coping unwittingly with the changes in food production and food security over this period. They must not panic now as they prepare for changes in the near and mid term in order at the least to maintain current food security or even to enhance it.

The future is arriving earlier than expected: 2020 is the new 2050

For the past decade or so, the public, which includes everyone, even scientists and policy-makers, has been informed by science media reporting about the buildup of greenhouse gases (GHGs) in the atmosphere. One of many effects of this buildup is global warming, which will increasingly intensify if societies continue on a "business as usual" path choosing not to alter their patterns of energy consumption and land use. Such scientific findings continue to serve as an early warning about foreseeable changes in the global climate patterns and the impacts of those changes on ecosystems and societies.

Based on the available data, scientists have developed scenarios that, in general, have focused on those climate changes and their impacts that might plausibly be expected to occur in 2050 or 2100, if all of the science is proven correct. While such processes of change are relatively well understood with regard to the state of climate science today, many of the rates of these processes of change are barely discernible over short time frames to the naked eye—and sometimes even to the instruments that measure such changes.

Reports are now coming in from scientists and are being repeated in the media worldwide that the rates of change for a wide range of ecological and social climate-impact factors are actually faster than had been predicted just a few years ago. As an example, the most visible rate of environmental change has to do with the accelerated disappearance of ice cover in the Arctic. Using sophisticated computer models, scientists had projected a certain percentage loss in sea ice cover in the Arctic by the year 2020; however, the disappearance of sea ice, based on actual measurements, had already reached those projected levels by 2007 — 13 years earlier!

The rapidity of the Arctic meltdown (and that of the Greenland ice cover as well) has sparked concern about rates of change in various ecosystems from the equator to the poles. Around the world, levels and impacts of warming that had been projected to arise many, many decades into the future are emerging now before our eyes. In other words, "the future is arriving earlier than expected." Such indications necessitate the shifting forward of consideration of the timeline suggested by climate change impact scenarios. This might help to show how quick the impacts might become visible in highly exposed climate- and water-sensitive sectors like agriculture.

The scenarios for 2100, while interesting to planners at some level, are of much less concern to most decision makers than are scenarios more proximate to our contemporary time of life and governance. If science is going to be relevant to most policy-makers today, then its projections must also include time scales that are far closer to the present than those a century away. Therefore, 2020, in the minds of those who are concerned with societal responses to a "dangerous" climate change and in light of accelerating rates of change, must be seen as the new 2050. Not only does 2020 become the new 2050, but the impacts projected for 2100, for example, may now plausibly arrive as early as 2050. Clearly, the climate is changing, and apparently far faster than we had expected.

The problem is that because the physical and ecological mechanisms involved in these processes continually seem to translate to shorter and shorter timeframes for what once were distantly projected impacts, these accelerated environmental changes will continue to create a major dilemma in thinking about and acting on these impacts, since both the physical and ecological rates of change will occur far faster than the rates at which institutional bureaucracies are designed to cope effectively. A further problem is that because the focus of the past decades has been on adapting and mitigating to future impacts, the concept of prevention seems to have been abandoned.

A RICE FARMER IN THE FLOOD PLAINS OF BANGLADESH

Smallholder subsistence farmers in Bangladesh depend on temporary transient livelihood activities after natural disasters. The rice field often turns into fish ponds.

3 ADAPTATION AND MITIGATION

DEFINITIONS

The "Four Laws of Ecology"

Over thirty-five years ago, American ecologist Commoner (1971) proposed Four Laws of Ecology in his book entitled "The Closing Circle". According to Commoner, "an effort has been made to develop this view [i.e., the laws] from available facts, through logical relations, into a set of comprehensive generalizations. In other words, the effort has been scientific" (p. 42).

These general observations about nature, proffered as laws, were proposed before global warming had become generally recognized as a major problem for society, for climate and the environment. Commoner's laws have, however, proven useful when discussing adaptation and mitigation strategies to cope with climate change and food security in a sustainability context:

1st Law ... *Everything is Connected to Everything Else.*
"The system is stabilized by its dynamic self-compensating properties; these
    same properties, if overstressed, can lead to a dramatic collapse" (p. 35).

2nd Law ... *Everything Must Go Somewhere.*
"One of the chief reasons for the present environmental crisis is that great
    amounts of materials have been extracted from the earth, converted
    into new forms, and discharged into the environment without taking
    into account that everything has to go somewhere" (p. 37).

3rd Law ... *Nature Knows Best.*
"The third law of ecology holds that any major man-made change in a
    natural system is likely to be detrimental to that system" (p. 37).

4th Law ... *There Is No Such Thing as a Free Lunch.*
"In ecology, as in economics, the law is intended to warn that every gain
    is won at some loss" (p. 42).

It is important that policy-makers at all levels of government keep in mind each of these "laws," as they search for, identify, develop and implement adaptation strategies for coping with the impacts of climate change on food security and implementing mitigation strategies to reduce greenhouse gas emissions. They serve as reminders of the important role of ecosystems not only in the health and well-being of societies but also on the health and wellbeing of other ecosystems on which those ecosystems depend. In a search for effective adaptation and mitigation strategies to enhance food security and produce bioenergy, the four laws can serve as educational and instructive guidelines to policy.

Food Security and the "Four Laws of Ecology"

When it comes to food concerns, fostering either food security or reducing food insecurity requires serious consideration of each of the four laws of ecology. For example, increasing biofuel production may require removing land from food crop production, which causes food prices in the marketplace to rise, the nutritional status of at-risk populations to decrease, and so forth [Ecology Law 1]. Furthermore, if large dams or irrigation systems are built to increase cash crop production capacity, people are forced to migrate. They may find new lands to cultivate the crops they have traditionally grown, but these new lands, because they are likely to be less fertile (increasingly marginal) than the lands they had been forced to abandon, will produce lower yields [Ecology Law 2].

Ecology Law 3, nature knows best, is well-illustrated by the previous scenario: People are often encouraged or forced to cultivate marginal lands, which are defined as lands that are not suitable for sustainable agriculture because of poor soils, inhospitable terrain, erratic precipitation, etc. Around the globe – and especially in the developing world – pressures to move onto new lands to grow food are increasing, as are pressures for both export and population demands and the marginalization of the poor.

The 4th Law – there is no 'free lunch' – is perhaps the easiest to illustrate. Changes in the ways societies choose to interact with the natural environment often produce winners and losers, relatively speaking. Large-scale cash crop and export-oriented irrigation schemes, for example, are usually implemented in areas with fertile soils, displacing local inhabitants and their traditional ways of coping with their harsh environments. Globalization has also put considerable pressure on local producers because of the cheaper prices for some imports, even though the many drawbacks of an inappropriate reliance on imports --- from concerns over energy consumption in transport to the shuttering of locally owned establishments --- are well-documented. Many such examples exist of situations where policy-makers believed they could "change one thing" but later learned that that one change led to a host of unintended consequences that proved more costly than the benefits they had gained from their policy decisions.

Adaptation

The IPCC's officially used operational definition for *adaptation* is as follows:

> *Adaptation* - *Adjustment in natural or human systems in response to actual or expected climatic stimuli or their effects, which moderates harm or exploit beneficial opportunities. Various types of adaptation can be distinguished, including anticipatory and reactive adaptation, private and public adaptation, and autonomous and planned adaptation (IPCC, 2001).*

Adaptation as a response to change must be appropriate to specific hazards or threats in a given period of time; in the same way, an effective adaptation to a real or perceived change in local climate could, over time, become inappropriate as circumstances changes. Importantly, however, some responses to change must be understood as reactive from the outset, because they were based on little forethought or analysis. Mal-adaptation refers to changes in the behavior of an organism (or a society) that prove counterproductive with regard to desired outcomes.

The word "adaptation," as important as it is to climate policy-makers and researchers, is, however, not defined in the same way by all who use it. Furthermore, for the sake of improved communication and understanding across disciplines and cultures, more people must become aware that several words, including (but not limited to) acclimatization, alteration, accommodation, modification, adjustment, are used as synonyms for adaptation. People must also be aware that these alternate terms, while synonymous in some respects to adaptation, do generate different understandings of what is happening in the name of adaptation. To understand the word "adaptation", Table 3.1 provides a few illustrative

TABLE 3.1

Selected examples of planned adaptation in the water and agriculture sector (IPCC, 2007b)

ADAPTATION OPTION/ STRATEGY	UNDERLYING POLICY FRAMEWORK	CONSTRAINTS	OPPORTUNITIES
Expansion of rainwater harvesting; water storage and conservation techniques; water reuse; desalination; water-use and irrigation efficiency	National water policies and integrated water resources management; water-related hazards management	Financial, human resources and physical barriers	Integrated water resources management; synergies with other sectors
Adjustment of planting dates and crop variety; crop relocation; improved land management, e.g. erosion control and soil protection through tree planting	R&D policies; institutional reform; land tenure and land reform; training; capacity building; crop insurance; financial incentives, e.g. subsidies and tax credits	Technological and financial constraints	Access to new varieties; markets; longer growing season in higher latitudes; revenues from 'new' products

examples of planned adaptation options, underlying policy frameworks, constraints and opportunities in the water and agriculture sector (IPCC, 2007b). They have a direct relevance to food security.

Mitigation

The 2nd Law of Ecology – everything must go somewhere – relates directly to the awareness that emitting greenhouse gases into the atmosphere in increasing quantities ad infinitum will have a major visible effect in the not too distant future on climate, ecosystems and societies. Adverse signs are appearing around the globe that strongly suggest that rates of change are occurring faster than scientists have been anticipating: most glaciers around the globe are melting, sea level is rising, warm temperature ecosystems are

TABLE 3.2

Key mitigation technologies and practices in agriculture and forestry, policies and measures, constraints and opportunities (IPCC, 2007b)

KEY MITIGATION TECHNOLOGIES AND PRACTICES	POLICIES, MEASURES AND INSTRUMENTS SHOWN TO BE ENVIRONMENTALLY EFFECTIVE	KEY CONSTRAINTS	KEY OPPORTUNITIES
Improved crop and grazing land management to increase soil carbon storage; restoration of cultivated peaty soils and degraded lands; improved rice cultivation; techniques and livestock and manure management to reduce CH_4 emissions; improved nitrogen fertilizer application techniques to reduce N_2O emissions; dedicated energy crops to replace fossil fuel use; improved energy efficiency; improvements of crop yields	Financial incentives and regulations for improved land management; maintaining soil carbon content; efficient use of fertilizers and irrigation		May encourage synergy with sustainable development and with reducing vulnerability to climate change, thereby overcoming barriers to implementation
Afforestation; reforestation; forest management; reduced deforestation; harvested wood product management; use of forestry products for bioenergy to replace fossil fuel use; tree species improvement to increase biomass productivity and carbon sequestration; improved remote sensing technologies for analysis of vegetation/soil carbon sequestration potential and mapping land-use change; Landfill management and monitoring	Financial incentives (national and international) to increase forest area, to reduce deforestation and to maintain and manage forests; land-use regulation and enforcement	Constraints include lack of investment capital and land tenure issues	Can foster poverty alleviation

moving upslope into higher altitudes and the areal coverage of Arctic sea ice is rapidly decreasing. As scientists have learned in recent decades where those greenhouse gases are going and what they are doing to the earth's atmosphere and oceans, governments have started to seek ways both to reduce their sources and to increase the sinks for those gases and to prepare for foreseeable adverse impacts.

Mitigation refers to technological change and substitution that reduce energy resource inputs and emissions per unit of output. Although several social, economic and technological policies would also lead to an emissions reduction, for climate change mitigation encompasses implementing policies to reduce greenhouse gas emissions and to enhance sinks. Table 3.2 provides selected examples of mitigation technologies, policies and measures as well as constraints and opportunities for agriculture and forests as outlined in the IPCC (2007b) Synthesis Report. Box 2 provides details of GHG emission and mitigation potential in food and agriculture sector.

A Danish government action program (DANIDA, 2005) defined mitigation in the same words as those of the IPCC: "[mitigation] is an intervention to reduce human-caused net emissions of greenhouse gases." Its report suggested some obvious measures that governments could pursue for mitigation:

- Reduction (at the source) of the use of fossil fuels (clean coal technology, renewable energies)
- Capture of methane from landfills and rice paddies
- Creation of sinks for storing carbon through natural resource management (carbon sequestration) [e.g. reducing tropical deforestation and increasing tree planting] [www.netpublikationer.dk/UM/5736/html/entire_publication.htm]

Mitigation policies, which require identifying effective ways to reduce the amount of greenhouse gases produced and released into the atmosphere, are the first and foremost line of defense for reducing emissions before the worst consequences of global warming are allowed to occur. Although mitigation is the preferred path, it is also perhaps the most difficult to achieve in a way that would have positive global results in a short time. One reason is that implementation of the many suggested mitigation techniques (e.g. transfer of clean technologies, switch to alternative sources of energy (including nuclear), capture and sequestration of carbon and other greenhouse gases such as methane, reduction of fertilizer use, more efficient

BOX 2

AGRICULTURE HAS POTENTIAL FOR CRUCIAL EARLY ACTION ON MITIGATION

The land area which is suitable for the production of food, feed, fuel, wood and other products provides a massive carbon store, but is also a source of GHG emissions. The specific aspects and options of GHG emission reductions and enhancing sinks in agriculture and forestry have the potential to mitigate GHGs in food and agriculture.

The Agriculture, Forestry and Other Land Use (AFOLU) sector is responsible for about one third of global anthropogenic GHG emissions. Land use is responsible for 17 percent of the emissions, mainly from deforestation, and agriculture contributes about 14 percent. There is an intimate connection between the different land use sectors, and in many areas agriculture is the main driver of deforestation, leading to GHG emissions.

The forest biophysical mitigation potential was estimated to be 5 380 Mt CO_2/yr on average up until 2050 (IPCC, 2001) and agriculture provides a technical mitigation potential of 5 500 to 6 000 MtCO_2-eq/yr by 2030 (IPCC, 2007d). Different forestry and agricultural practices and measures exist which provide mitigation opportunities.

The emissions caused by agriculture can be reduced by more efficiently managing the carbon and nitrogen flows. This can be induced through a change in management practices. For example it is possible to reduce the emissions of CH_4 from livestock by increasing the feed use efficiency or from crop production by adopting practices that enhances Nitrogen use efficiency by crops decreasing the emission of N_2O. The emission reduction potential differs between areas and sectors.

GHG emissions can be avoided or displaced. Fossil fuel energy can in some cases be replaced by bioenergy from wood, agricultural feed stocks and residues and/or the energy efficiency in agricultural sector can be improved. Agricultural mitigation measures often have synergy with sustainable development policies, and many explicitly influence social, economic, and environmental aspects of sustainability. Sustainability criteria need to be applied to ensure sustainable soil and water management and the protection of high biodiversity and nature reserve areas.

Agriculture and forestry have the technical potential for climate change mitigation. The overall challenge is to transform this technical mitigation potential into practice. We have, on a research basis, suitable technologies and farming practices, measurement technologies and experiments with payments for ecosystem services. Approaches to carbon sequestration in smallholder contexts can therefore be developed. Agriculture mitigation practices, such as crop and grazing land management, agroforestry and restoring cultivated organic soils generate high co-benefits for the smallholders, such as raise in productivity, household food security, and increased resilience and ecosystem services. For mitigation activities to become effective a comprehensive landscape approach is necessary.

However, the challenge is to design financing mechanisms for the remuneration of environmental services in the smallholder agriculture. These mechanisms need to provide an incentive for providing and safeguarding ecosystem services such as watershed protection, carbon sequestration and biodiversity provision. For smallholders to be able to participate and benefit from financial rewards and adopt mitigation practices, mechanisms need to be designed which cover up-front investment costs. Institutional set ups are required to aggregate the mitigation reductions across smallholders in order to reduce monitoring and transaction costs.

use of water resources in agriculture and in urban centers) would depend on both the decisions and the will of national policy-makers in industrialized and developing countries alike. Notably, some proposed mitigative tactics (e.g. mirrors in space, iron particles in the ocean, application of reflective particulates in the stratosphere) center on massive planetary engineering schemes that border on science fiction and that could, in turn, result in unintended and even dire consequences.

Numerous plans that include non-engineering solutions have also been proposed by various governmental, intergovernmental and non-governmental organizations to reduce greenhouse gas emissions (specifically CO_2 emissions). Carbon trading, for instance, would be a market-based system established between those states that emit greenhouse gases above an allowable country level and those that emit below the amount they are allowed to emit.

Researchers at the World Resources Institute, recently published an article about how to enhance climate change mitigation opportunities in the U.S. agricultural sector that provided useful information and policy options for coping with the emissions of nitrous oxide and methane. The article suggests ways that managers of agricultural operations can reduce their greenhouse gas emissions (WRI, 2007). Policy implications are also noted. The following paragraphs are directly from this article (http://pdf.wri.org/ agricultureandghgmitigation.pdf):

Nitrous oxide

N_2O comes from two main sources—livestock manure and chemical fertilizers. When bacteria interact with ammonia, N_2O is released. Therefore, to reduce N_2O emissions, farmers must decrease either direct emissions of N_2O or the amount of ammonia produced during normal agricultural processes. In dairy and cattle operations, large amounts of ammonia are produced when urea and livestock manure break down in water or slurry. Even greater emissions come from field operations, with the applications of nitrogen fertilizer and related cropping practices.

Since fertilizer is responsible for large amounts of agricultural sector N_2O emissions, farmers can choose management practices that lead to appropriate fertilizer application rates. N_2O emissions [can be decreased] by avoiding costly fertilizer over-application.

Methane

The agricultural sector is, for example, the second largest contributor of CH_4 in the United States, with approximately 70 percent of agricultural CH_4 emissions coming from enteric fermentation, 25 percent from the decomposition of manure, and 5 percent from rice cultivation.10 Enteric fermentation is a natural process that occurs in the digestive systems of animals such as cattle, sheep, and goats. As much as 7 percent of an animal's *feed can be lost as CH_4, so feedlot operators who increase animal digestive efficiency will save feed costs and decrease methane emissions. Options for increasing efficiency include increasing the daily percentage of highly digestible feed and correcting nutrient deficiencies in livestock diets.*

Manure stored in central tanks or lagoons also releases CH_4 during anaerobic decomposition. However, new technologies now make it possible for this excess CH_4 to be captured and either used directly or sold as energy. Capturing the released CH_4 and using it for energy effectively reduces GHG emissions, while also helping to meet on-farm energy needs and reduce electricity costs.

Finally, rice production is responsible for CH_4 emissions from agriculture. These emissions are generated through the cultivation of wet rice, which promotes the anaerobic decomposition of plant wastes that remain after harvest. Reductions in CH_4 emissions can be achieved by using different rice cultivars, improving water management practices, and efficient use of inorganic fertilizers.

Carbon dioxide

A majority of these emissions [in agriculture] are related to land-use change (i.e., deforestation), diesel fuel use, and energy used for irrigation and drying of grain. Increasing cultivation efficiency by moving to low-or zero-tillage crop management practices, using more energy-efficient machinery, or reducing energy demand will reduce these direct CO_2 emissions. While agriculture emits only small amounts of CO_2, it has the capacity to store carbon in plant material and soils. However, this ability to store carbon is limited. Best management practices include conservation tillage, nutrient management, rotational grazing and improved forage management, use of cropping rotations and cover crops, and the establishment of riparian buffers. For farmers to benefit financially from providing carbon offsets using these best management practices, policy-makers will need to develop systems for inventorying and monitoring soil carbon in agricultural lands.

Trade-offs

Inevitably there will be conservation practices that benefit one natural resource while harming another. Leaving water on land under rice cultivation to promote wildlife habitat, for example, can increase wetland *acreage and enhance wildlife benefits, but can also accelerate the generation of CH_4. An example with positive benefits is where reduced nitrogen fertilizer applications improves water quality and also reduces N_2O emissions. Similarly, riparian buffers enhance wildlife habitat, improve water quality, and increase carbon storage.*

Conservation practices may also have varying effects on different GHGs. For instance, capturing CH_4 from livestock manure and urine involves storing the material. Storage reduces the exposure of the urine and manure to oxygen, thus decreasing the release of N_2O. This illustrates how one conservation practice can simultaneously lead to reductions in two GHGs. Thus, estimating environmental outcomes from conservation practices is important.

In Pursuit of Resilient Adaptation to climate change and its impacts

"Resilient Adaptation" is a hybrid concept that merges the best of the suggested practices of resilience and of adaptation in the face of potential hazards and threats from climate change. It includes a safety net or way out of strategies that may, after a while, prove to have been mal-adaptations. It also includes a recovery mechanism that has a degree of flexibility in the face of uncertain future, scientific model-based findings notwithstanding. The concept of resilient adaptation is borrowed from the field of psychotherapy. The editor of a book on the topic suggested "resiliency is operationally defined…as a dynamic developmental process reflecting evidence of positive adaptation despite significant life adversity" (Luthar, 2003). The notion of Resilient Adaptation can be applied to societal as well as individual well-being in terms of climate change assessments on adaptation and mitigation.

SWOC/T assessment of scenarios for adaptation

SWOC/T assessments are used to evaluate the Strengths, Weaknesses, Opportunities and Constraints (or <u>T</u>hreats) of an organization, process or plan. They can also be used as educational tools to assess the prospects and potential pitfalls of strategic responses a government might pursue to counter the adverse impacts of climate change or to derive value from the transformations in the environment that a change in climate might cause. In an open forum, a SWOC/T approach can also help tease out those not-so-obvious aspects of a policy response to climate change's influences on a country's food security. In addition, exposing weaknesses can be useful in a government's preparations for or avoidance of the adverse side effects of a strategy's implementation. In the same vein, identifying both obvious and not-so-obvious constraints is the first step in identifying pathways to remove or overcome them. Although SWOC/T assessments can be valuable learning tools, they will not in and of themselves yield designs for strategic plans to cope with climate change's impacts on food security.

Scenarios

The creation of scenarios (for example, "Forecasting by Analogy" noted above) is a popular approach to attempt to gain a glimpse of the future, at least the near-term future. Scenarios can help decision makers create contingency plans for possible futures based on past experience. Surprises are to be expected, of course, even though the form they will take may not be known, but scenarios, overall, can be quite useful for hypothesizing about a wide range of potential impacts of a changing climate. As an example, decreases in the area covered by snow and ice in the Arctic are predicted as the earth's atmosphere warms; however, the rates of melting and disappearance of sea ice are now happening much faster than scientists had originally estimated. This means that increased rates of warming can be expected because, unlike snow and ice that reflect a large proportion of solar radiation back into space, ocean water absorbs incoming radiation, forming a positive feedback loop that will result in increasing temperatures.

Scenarios are like contingency plans: They have a limited shelf-life. As an example, 13 months before Hurricane Katrina made landfall in 2005 along the coast of the Gulf of Mexico, destroying the US coastal city of New Orleans, local through national government officials had gathered in the region for an exercise on how to respond to the impacts of a hypothetical Category 3 tropical storm. They called the hypothetical storm Hurricane Pam. Unclear even until now is the extent to which lessons that were allegedly "learned" during the Hurricane Pam scenario exercise were actually followed The US government's initial response (or lack thereof) in the early days of Hurricane Katrina suggests that the Hurricane Pam scenario had little influence on decision making when it was confronted by a real disaster. It appears that the Hurricane Pam exercise had become a distant memory to planners by the time Hurricane Katrina had formed in the Gulf of Mexico.

Nevertheless, scenarios are useful heuristic devices that provide insights to users about the potential demands of structures and functions of institutions and processes. They highlight the potential needs of a society to reduce vulnerability to threats and to increase resilience. Because of their relatively short shelf-life and because societies are constantly changing, however, scenarios need to be revisited, critically reviewed and updated periodically at regular intervals.

PRIORITY SETTING
Foreseeability and the Precautionary Principle

Foreseeability is a legal concept used to determine negligence. "In the Law of Negligence, the foreseeability aspect of proximate cause [primary cause of injury] is established by proof that the actor, as a person of ordinary intelligence and circumspection, should reasonably have foreseen that his or her negligent act would imperil others..." [http://legal-dictionary.the freedictionary.com/foreseeability]. Foreseeability has positive value for its use in terms of climate change.

Foreseeability differs from the concepts of forecast or predictability because it neither depends on nor implies any quantitative description of probability of occurrence. It suggests, for example, that a reasonable person can conclude that certain agricultural practices in certain types of ecosystems, in the absence of any action to change them, will have knowable adverse impacts on environmental quality. Those adverse impacts can lead to such degrading processes as soil erosion, deforestation, fertilizer and pesticide overuse, excessive water diversions, salination of irrigated soils, mechanization of land-clearing activities in increasingly marginal areas, excessive wood gathering for charcoal production for various reasons, and so forth.

These are some of the impacts that occur under today's climate conditions. As the climate warms, however, policy-makers must be prepared to identify and respond to early warning signs of the subtle changes in the local characteristics of their own specific climates. Early warning systems are necessary to alert them to such changes. In addition, they must become increasingly risk-averse in the face of an unknown future. In other words, they must consider using the "Precautionary Principle" when making decisions that might have consequences for food security. The "Precautionary Principle" is a political decision-making approach that emphasizes that a lack of full scientific certainty should not be used as a reason for communities and governments to postpone action to prevent serious and irreversible environmental damage (WLVC, 2003 [http://www.ilec.or.jp/eg/wlv/complete/wlv_c_english.PDF]).

A wide range of climate and climate-related impacts on society can be analyzed through both foreseeability and the "Precautionary Principle." By looking at how climate impacts in recent times have adversely or positively affected food security, for example, governments and humanitarian agencies might effectively determine what characteristics of drought had actually been foreseeable and apply the "Precautionary Principle" the next time those characteristics are identified to mitigate the impacts of future, similar threats. Numerous examples of when existing, reliable information was not used as an impending climate-related food shortage approached and a full-blown food crisis emerged can be cited (Glantz and Cullen, 2003).

Knowable surprises: surprises that shouldn't be surprising

Arguably, most climate and climate-related surprises are knowable at some level of awareness, especially as scenarios and historical re-enactments better enable the identification of many potential surprises. Myers and Kent (1995) noted:

> It might seem fruitless to speculate about seemingly unknown problems in the environmental field. But recall that at the time of the first major international conference on the environment in Stockholm in 1972 [UN Conference on the Human Environment], there was next to no mention of what have now become established as front-rank problems: global warming, acid rain and tropical deforestation.

To this illustrative list of seemingly unknown or unimportant topics could be added, among others, coral reefs, mangroves, desertification and biodiversity.

A central constituent of any of the various definitions of "surprise" is the word "unexpected"; indeed, the concept of the unanticipated is, for most people, fundamental to the characterization of an event as surprising. In this way, surprise relates to the "3rd Law of Ecology" (that Nature knows best) in that societies must respect and accept the fact that scientists are as yet unable to forecast with a desired levels of accuracy the variations and changes of climate and weather on time scales of interest to societies and their leaders. Not surprisingly, therefore, events will befall societies that could not have been anticipated, given our current state of knowledge of the climate system. For example, in 2004 a hurricane, for the first time in history, appeared in the South Atlantic and made landfall on the Brazilian coast. This event was truly surprising.

When trying to forecast surprises to prepare for them, problems often arise because of this reality that the exact timing, intensity, location or duration of events can often not be known or knowable. But climate and weather surprises are not always only physical; they can also arise as a result of perceived impacts. In fact, human perception is a key facet of how societies or groups within societies view the concept of "surprising." It should not be surprising, for example, that as the temperature of the atmosphere increases, some plants will fare well while others will not because, although the exact responses of specific plants remain unknown to researchers, the fact that flora is pretty much temperature and rainfall dependent is elemental biology. Undeniably, many signs have already emerged indicating shifts in the behaviors of a range of plant species with the already-warmed climate. The question, then, is whether or not this constitutes a *knowable surprise*?

Although the phrase sounds a bit contradictory, the fact is that there are knowable surprises, especially if the common usage of the word 'surprise' as opposed to its strict definition is considered. People who live in certain areas around the globe know that droughts are a part of their climate regime, for example. The fact is that drought will come with some frequency, although the exact onset of the next drought and its duration might be unknowable in advance. Similarly, in some areas where locust swarms appear from time to time, governments expect them, though they may still be surprised by the timing of a return, the magnitude and duration of an episode, or the extent of damage to the agricultural sector. The same can be said of flood-or fire-prone areas. The point is that there will always be unknowable aspects to expected events – knowable surprises.

Invisible boundaries: traditional conflicts involving agriculture

Agriculture has for centuries if not millennia been directly and indirectly involved in various controversies and disagreements (conflicts) related to food security. Often, these controversies are posed as dichotomies, as illustrated by the following non-exhaustive list of traditional agricultural conflicts:

- Agriculture vs. environment
- Intensive vs. extensive agriculture

- Food self sufficiency vs. exports
- Cash crop vs. food crop
- Food crop vs. biofuel crop
- Crops for export vs. crops for domestic consumption
- Globalization vs. localization of agriculture
- Global food security vs. household food security
- Government priorities vs. farmers' (patoralists') priorities
- Open rangelands vs. feedlots
- Trans-border migration for earnings Vs. trans-border migration to sustain livelihoods (eg. herders)
- Cultivated areas vs. rangelands
- Irrigated agriculture vs. rainfed agriculture
- Small scale irrigation vs. large scale irrigation
- Agricultural practices vs. water quality
- Virtual: water for export
- Urban vs. rural food prices
- Agricultural pressures vs preserved areas
- Cultivated areas vs forested areas
- Mangroves and agriculture farms vs. shrimp farms
- Large-scale mechanized fishery vs. small scale fishers
- Existing land use pressure vs. additional pressure from temporary and permanent refugees
- Inorganic agriculture vs. organic agriculture
- Mechanized agriculture vs. small scale indigenous agriculture with traditional draught power
- Agriculture intensification vs. biodiversity conservation
- Biofuels promotion vs. biodiversity conservation
- Genetically Modified (GM) crops vs. traditional crops
- Agricultural failure in conflict zones

Although each of these conflicts/controversies are posed here as simplistic "either/or," zero-sum pairs, the reality is that they all exist in multifaceted interrelationships involving societies, climates, economies, etc. If stakeholders and political gatekeepers can consider how these controversies and conflicts will be affected by global warming, however, win-win solutions could become possible that bring opposing sides together to overcome the challenges that will be generated by warming.

Invisible boundaries: water-related traditional conflicts and controversies

Similar types of conflicts and controversies can be identified for water. A suggestive list of some of them includes the following:

- Upstream practices vs. downstream practices
- Surface water vs. groundwater
- Rain water harvesting vs. installation of deep tube wells
- Natural flows vs. Reservoirs and dams
- Societal vs. ecosystem use
- Water rights vs. water responsibilities
- Water transfers from surplus to deficit regions
- Irrigation vs. rainfed agriculture
- Virtual water (in-country; exported water)
- Water for agriculture vs. water for urban areas
- Water for agriculture vs. water for eco-tourisms
- Water for agriculture vs. water for industry
- High Yielding varieties vs. traditional crops
- Drainage water vs. storage facilities

A "heads up" warning about how global warming might influence the invisible "frontlines" of these controversies and conflicts can be a first step towards the development of issue-specific anticipatory resilient adaptation strategies. Each of the conflicts or controversies in the list above has generated a considerable body of literature, both peer-reviewed articles and grey literature in the form of government and non-governmental reports.

Given the specter of climate change at local, national and regional levels, populations, disease vectors, animals, fish populations, ecosystems, rainfall patterns, etc. can be expected to shift in time and space. Known patterns of interaction, either peaceful or conflict-laden, can also be expected to change. Such changes, however, if anticipated correctly, can lead to future cooperation as opposed to continuation of existing conflicts. New relationships can be forged. Indeed, the more researchers and policy-makers know about the local to regional changes expected to accompany global warming, the better their opportunities will be to manage potential cooperation and minimize potential or defuse existing conflicts. The specter of continued climate change throughout the rest of the twenty-first century could, in the end, foster a time for immediate, urgent conciliation between competing and conflicting forces and interests.

Given the uncertainties surrounding the science and the potential uses of scientific information in decision-making, making explicit these and other agriculture-related controversies provides an excellent opportunity to pursue disaster-related diplomacy (in this case, disaster-avoidance diplomacy) to shape compromises as protagonists will face the same pressures and uncertain futures as a result of global warming (www.disasterdiplomacy.org).

Invisible Boundaries: Food, energy and climate

Food, energy and climate. For the first time in history, these three are closely linked. Without an understanding of this new reality, countries and the international community lack for the most fundamental policy decisions – decisions that affect access to food for millions of people. (FAO, 2008)

The high level conference on "World Food Security: Challenges of Climate Change and Bioenergy" has recognized the importance to address the challenges and opportunities posed by biofuels, in view of the world's need for food security, energy and sustainable development. The governments have highlighted the importance of in-depth studies to ensure that production and use of biofuels are sustainable in accordance with the three pillars of sustainable development. Biofuel development must also takes into account the need to achieve and maintain global food security. To foster a coherent, effective and results-oriented international dialogue on biofuels in the context of food security and sustainable development governments need to understand the linkages and controversies surrounding food and fuel. The following list highlights some of the controversies that exist over biofuels.

- Food vs. Fuel
  Corn (maize) is used for much of the ethanol production in the world, and the US, the European Union and other governments have mandated that a certain percentage of fuel include ethanol. As a result, many of the stakeholders in the corn production, marketing and sales chain have reaped financial benefits in sales for biofuel production rather than food production.
- Fossil Fuels vs. Biofuels
  Some biofuels produce less carbon dioxide to the atmosphere than others. Corn used in ethanol production was once believed

to produce less CO_2, but now science suggests that more CO_2 is omitted if both the production and the use of corn-based ethanol are accounted for. On the other hand, Brazil argues—and science at present supports—which biofuels produced from sugar cane, clearly emits less CO_2 than fossil fuels.

- Biofuels vs. "Biofools"
  While some in the bioenergy business tout that biofuels can help lower energy prices as well as dependence on foreign oil imports, others consider them foolish, arguing that biofuels, even by the most generous estimates, will replace only a few percentage points of a country's total energy consumption. Critics of "biofuels as panacea" see them more as a temporary band-aid than a real solution to the larger problems of fossil fuel consumption.

- Cash Crops vs. Food Crops
  A constant battle is fought between those who want to put arable land (rainfed and irrigated) into the cultivation of cash crops for sale to export markets and those who want to increase food production for domestic consumption. To the list of traditional cash crops must now be added crops that were once grown solely for food consumption but are now mainly diverted for use as feedstock for biofuels.

- High Energy Prices vs. High Food Prices
  Because high energy prices are a major cause of the high cost of food in marketplaces worldwide, a debate currently exists over whether biofuel production increases energy or food prices.

- Agricultural Land vs. Marginal Land
  Those pursuing the development of bioenergy contend that only unused or marginal lands will be used for biofuel production; no land is to be taken away from food production, they claim. That has not been the case for corn or soybeans in the USA and elsewhere, however, as many thousands of acres of productive farmland has been diverted in recent years to produce crops for feedstock and not for foodstock. In addition, some countries are felling trees in once-protected rainforests to develop palm oil plantations for biofuels.

- Affluence vs. Poverty
  Some countries are apparently considering securing large tracts of land in developing areas in order to grow food for their domestic

©FAO/G. Bizzarri

SARDINE FISHING IN THE ATLANTIC OCEAN, EL JADIDA, MOROCCO
The management strategy for fisheries for a given place must reflect a level of fishing effort given the numerous uncertainties that can surround the exploitation of living marine resources.

markets because they do not have enough arable land within their borders to meet domestic needs. This is a major ethical issue because, for one example, poverty-stricken, food insecure Africans will soon be growing food for affluent populations, which, especially in Asia, are rapidly growing. What this means is that African subsistence farmers are likely to end up as landless migrants laboring on farms that produce food for other countries.

- Food Security vs. Food Insecurity
  The increasing expansion of biofuel production on land traditionally used to produce food will likely generate food insecurity, even in places where it had not existed before. While biofuels can generate foreign exchange that can in theory be used for development purposes, those funds are often diverted to other pet projects of a country's leaders of politically connected organizations.

Agriculture-related invisible boundaries are shifting

Controversies and conflicts are dynamic between groups with different competing perceptions about how best to use land or ocean resources. Part of this dynamic often results from government policies. Governments, for example, may encourage cultivators to farm rangelands, displacing pastoralists. But part of this dynamic is likely to be climate-related: During extended drought periods, pastoralists, on the other hand, may be forced to abandon drought-desiccated rangelands and migrate towards wetter cultivated areas, perhaps encroaching into some of the former rangelands that had been overtaken by farmers in earlier, wetter periods. In other words, there can be advances by one side in the controversy and retreats by the other, and vice versa. In another way, one side of a controversy may superficially have "won" the conflict by, for example, dominating a particular swath of land, though in the long run that side may prove to be the biggest loser, having wantonly destroyed a mangrove forest to develop a poorly planned shrimp farm that ended up devastating the ecosystem upon which shrimp populations depended. In all cases, a result to be avoided of human interactions with the environment is one in which, in the long run, "the winner takes nothing."

Winning and losing in agriculture under a warmer atmosphere

Winning and losing, when applied to climate change, is a controversial topic that requires more clarity. If one were to inform a person in an arid area that there would be an increase in precipitation, at first that person might consider it a "win". However, there is no information about when or how that precipitation might be delivered. If it fell in downpours in one super-storm event, then that increase would not have been considered a win but would be a clear loss. The point is that there has been no attempt to systematically and specifically identify region by region what changes in the aspects of climate might be advantageous to a society and which ones would be harmful.

As noted earlier, government leaders do not usually make decisions based on global statistics and global averages. Agricultural production is a local affair, but trade, aid and comparative advantage make agriculture a global affair. Research suggests that some crops will do well in a somewhat warmer atmosphere, while others will not. Some locations are expected to do better in term of crop yields in a warmer climate, while others will do worse. There are still many unknown factors when it comes to speculation about crop production and crop yields under a warmer climate regime: the hydrologic cycle will intensify, all scientists seem to agree on that, but where, when, and how will that additional precipitation fall? We already know, for example, that crop production is on the rise in, of all unsuspected places, Greenland. So, from the perspective of the Government of Greenland, that is good news; an increase in food self-sufficiency (they can now grow broccoli). The bad news for the government, however, is that Greenland is shrinking in overall size as its ice cover melts.

One can easily argue that, under the current climate (given its average, variability and extremes), different countries, socioeconomic sectors and groups have had identifiable relative (comparative) advantages and disadvantages. This results from an interplay of climatic factors with unique sets of economic, social and political factors. Gains and losses at all levels of society will foreseeably result either from the local climate change itself or from the way that humans respond to that change. Some countries, sectors or groups may have the capability to respond (adapt) to climate change, turning this to their future advantage.

However, with regard to global warming, researchers talk about two phases: a transitional process and an end state. While climate changes in the near term may appear to some countries either as a benefit or a loss, over the long term they argue that there will be no winners. All will lose. Thus, it is also foreseeable that those who benefit in the near term might not fare as well, as the climate continues to warm. So, what might appear a benefit now may turn into a loss in the future, and vice versa. Policy makers must be aware of this possibility.

As noted in the IPCC 3rd Assessment (IPCC, 2001), many rainfed crops in Africa and in Latin America are at their limit of tolerance with regard to temperature. It suggested that productivity in these areas could decline up to 30 percent while productivity of corn in Europe, for example, could increase by 25 percent. Although the 4th Assessment Report states that in the mid- to high-latitude regions, a moderate warming of the climate would benefit crop and pasture yields, even just a slight warming will likely decrease yields in seasonally dry and low-latitude regions [NB: this IPCC projection was made with medium confidence (IPCC, 2007c)]. The point here is that there are knowns, unknowns, and uncertainties about how climate change might affect agricultural productivity, other things being equal, but in most cases other things are never equal.

Once again, the Four Laws of Ecology are relevant: warmer temperatures affect precipitation in time and space as well as evapo-transpiration rates, cloudiness, changed possibilities for pests and invasive species, changes in the characteristics of the seasons, and the need for and development of new technologies and techniques, and so forth. I would suggest that, in general, it is a bit too early to identify all the winners and losers in agriculture, livestock and fisheries, although new evidence of agriculture under a changing warmer climate is constantly emerging.

Participants in a 1990 climate impacts workshop "On assessing winners and losers in the context of global warming" preferred not to talk of winners and losers but to talk of the advantaged and disadvantaged. The former set of terms implied there was an end state in the evolution of human interactions with the changing climate (Glantz, 1990). Yet another, less confrontational, way to describe wins and losses for global warming would be to refer to the "preferential access to food and other resources".

Biofuels and early warning systems

In just a few years biofuels have tumbled from their position as the 'darlings' of development (lower cost energy, reduced CO_2 emissions, generation of sorely needed foreign exchange, an expansion of trade) to become a solution now collectively scrutinized by a growing number of observers as problematic both for the environment and for long-term development prospects. Today, any discussion that contains the word "biofuels" generates controversy. What seemed like a good idea with win-win consequences for environment and for society, producing energy from biological matter, has unleashed a whirlwind of accusations and finger-pointing, of point and counterpoint, on the benefits and pitfalls of biofuel production and use.

Recent, though post-facto (belated) analyses of biofuels production have raised questions from a climate impacts standpoint about their expected contribution to reducing greenhouse gas emissions. Corn, for example, has changed in the estimation of many from a good crop to a bad crop based on findings of its global warming potential (GWP) alone. Further study has shown that the process of manufacturing corn-based ethanol, as opposed to reducing GHG emissions as originally thought, actually contributes more to greenhouse gas emissions than the burning of most fossil fuels. But large tracts of land have been and continue to be leased for decades or more on the hope and prayer of biofuels' benefits to environment and society: the hope is that biofuel production will lead to prosperity and economic development and the prayer is that investment in land and labor for biofuel production will, on a plant by plant and a case by case basis, withstand a SWOC/T assessment conducted by an independent party.

In retrospect, would the questions elicited by applying the Precautionary Principle about the impacts of biofuels on the environment, society, and the economy and posed in advance of the rush to produce such fuels have revealed some of the late lessons that appear to be emerging despite the early optimism? Would an early warning assessment of biofuels have been of value? Had these conflicts been identified in advance, precautionary steps – a preliminary assessment of potential impacts in the form of a warning system, a feasibility study, or an impacts assessment, for example – could have been taken along the lines suggested by the "Precautionary Principle" before actions too difficult to stop were taken. What remains unclear is the degree to which biofuels will prove to have been a good supplement to the energy needs of countries.

Recently, the emergence of a new group of energy investors has muddled the issue even further. These are speculators and corporations who are entering the energy business in anticipation of sharp, quick gains on their financial investments in the conversion of biological matter into biofuels. Other governments are investing in biofuels to make money to enhance their economic development prospects, often encouraging the involvement of those energy speculators who are providing them with extremely unfavorable investment terms. The well-known truth is that countries in the twenty-first century require energy to function, and energy corporations are reaping enormous profits by setting the terms by which that need is being met. Even though the technologies to meet their capacity demands and the support of a majority of their constituents exist, alternative energies are still not pursued seriously by most governments in an all-out war on dirty greenhouse-gas-producing energy sources in favor of truly cleaner solar and wind energy (and even a serious all-out approach to conservation). Instead, many governments see tremendous potential in growing their own feedstock for biofuel production to relieve domestic pressures on their energy needs. What seems to be going on right now, in essence, is an energy version of a good old-fashioned high-school-style food fight, but instead of the school cafeteria, the battleground is Planet Earth, and instead of students hurling mashed potatoes and cherry pie are "brown-eyed," "blue-eyed," and now "green-eyed" energy entrepreneurs fighting for a larger share of the profits to be made in the energy sector, heedless of the fact that someone, someday will have to mop up the mess.

Biofuels have all the markings of a classic "boon to bust" phenomenon. As renowned engineering professor Henry Petroski once wrote, however, "hardly a history can be written that does not include the classic blunders, which more often than not signal new beginnings and new triumphs" (Petroski, 1992). He also suggested that "Failures in turn lead to greater safety margins and, hence, new periods of success". The image that comes to mind when contemplating today's energy quagmire is that of deckhands re-arranging the chairs on the Titanic in the minutes after it hit the iceberg. Instead of focusing on how best to save the passengers, the captain and his crew – by analogy, those in the energy business as well as myopic policy-makers – are busy rearranging the deck chairs to obtain a better view of the iceberg that caused the gash in the hull.

©FAO/A. Proto

A YOUNG GIRL LEADS HER DONKEY, HEAVILY LADEN WITH JERRICANS OF WATER, THROUGH THE DESERT IN SUDAN

The consequences of climate change are complex and far-reaching. Climate change will affect all water-related sectors, including drinking water, agriculture, ecosystems, navigation and hydropower.

4 WHAT TO DO AT THE NATIONAL LEVEL

TODAY'S "BEST PRACTICES" MAY NOT BE ENOUGH

There is a great deal of interest in – and perhaps reliance on – the concept of "best practices," which can provide a useful starting point for brainstorming to develop new climate change related-strategies. The UN Habitat website highlights its best practices database on improving the living environment and provides the following brief overview of the purpose of best practices (UN Habitat, 2007):

> This searchable database contains over 3 800 proven solutions from more than 140 countries to the common social, economic and environmental problems of an urbanizing world. It demonstrates the practical ways in which public, private and civil society sectors are working together to improve governance, eradicate poverty, provide access to shelter, land and basic services, protect the environment and support economic development. (www.bestpractices.org)

FAO's initiative on Technology for Agriculture (TECA) provides proven technologies (which are also relevant to climate change adaptation and mitigation) for smallholders and aims at improving access to information and knowledge about available proven technologies in order to enhance their adoption in agriculture, livestock, fisheries and forestry thereby contributing to food security, poverty alleviation and sustainable development. (http://www.fao.org/teca/)

In regard to climate change and the likelihood that future characteristics of global, regional or local climate will change in unknown ways, however, existing best practices should be viewed as providing a source of tactical responses (short-term) to a changing environment as opposed to an acceptance of untested strategic responses (longer term). The reason for this is because climate impacts and response mechanisms to them in the near-term future are likely to be similar to those of the recent past, barring any abrupt changes in the atmosphere's local to global climate characteristics. The characteristics of change and impacts in the future, on the other hand, are more uncertain.

Both short-term and long-term policy options, therefore, are the way ahead for the FAO and its partners. "Best practices" could provide a pathway to the near term future, given the high level of uncertainty that surrounds climate change impacts on societies and both managed and unmanaged ecosystems. For the long term, though different means by which to approach adaptation will be required, because of increasing uncertainties about the future and the absence of the not-yet-identified benefits of ongoing research and its usable findings.

"ORDINARY" KNOWLEDGE ABOUT FOOD SECURITY

Farmers and herders worldwide have relied on best practices for millennia as they came to understand them for growing food for consumption or barter or for raising livestock for the same reasons. They did so through trial and error, with one generation passing its success and failure stories on by word of mouth or by demonstration to successive generations. They did not go to school for their education on how to work the land or to manage its resources, nor did they calculate probabilities in a quantitative way. They watched the environment and developed intuition in reading cues about when, what, and where to plant, when to water different crops and when to harvest. They learned how to store and carry food over from one season to the next. Such education has similarities to a person in modern society who knows what to do when crossing a busy street. They are guided by rules such as "look both ways" that were taught to them, often informally, when they were children. Both experience and intuition reinforced those rules until they became second nature, enabling that person (and most people most of the time) to cross streets safely with little or no quantitative skills, information or calculations about velocity or laws of motion in hand. Such a scenario for street crossing in a big city is as likely true for people who have had formal education as for those who have not.

Policy-makers are drawn from civil society and are likely to rely on their own ordinary knowledge as well. Similarly, agricultural researchers have a responsibility to listen to the public and its views as reliable input based on ordinary knowledge for decisions about food security. But scientists have a further responsibility – to make clear the results of their research, correct misinterpretations of environmental cues and foster proper use of scientific indicators in ways that reinforce or calibrate "ordinary" knowledge (Lindblom and Cohen, 1979).

Regrettably, communication between scientists and the public has apparently been inadequate for a very long time. As H.G. Wells wrote over one hundred years ago (1904), "many of those scientific people understand the meaning of their own papers quite well. It is simply a defect of expression that raises the obstacle between us." (Wells, 1904). Today, given the relatively rapid changes underway in the climate system, ordinary knowledge will need to be supplemented by scientific knowledge in ways that laypeople understand. Fortunately, innovations such as wireless communication technologies are constantly being developed and becoming economically feasible for large sections of society. These technologies must be exploited to enhance, for example, communications among climate scientists, policy-makers and farmers/ herders that would enable a social discourse that would surpass the top-down strategies of the past in favor of the more equitable possibilities for action and understanding that emerge when voices from all stakeholders are heard. Increased communications would also enable meaningful lateral interactions between, for instance, illiterate successful farmers and herders who are empowered to teach other illiterate farmers and herders who are less successful.

"ONCE IS NOT ENOUGH"

Adaptation is an ongoing process, and developing an initial set of strategic adaptation responses to the potential impacts of climate change on food security is only the beginning of that process. The climate will continue to vary and change as will its impacts on ecosystems and the human activities that are dependent on them. Many of both the obvious and subtle changes witnessed so far have been similar to those of the relatively recent past. Because the scientific community does not yet know with a reliable degree of confidence how high global and local temperatures will rise throughout the twenty-first century, decision makers must maintain a degree of flexibility in the application of their adaptation strategies and tactics. The reality of this unknown suggests that considerable precaution must be taken in policy making for food security under global warming projections. Resilient adaptation as a response provides the necessary flexibility to cope over time with a changing climate.

A STEP BEYOND: MITIGATING THE IMPACTS OF ADAPTATION

In present discussions of climate change, adaptation rules the day. The general belief is that little can be done to prevent the coming impacts, so societies have no choice but to prepare for those impacts by developing adaptation measures. However, the societies have to look beyond reacting to climate change to the downstream impacts of their proposed adaptive policies and practices, as they too will generate their own impacts. Many examples illustrate how a lack of foresight when it comes to coping with hazards and disasters has led to other challenging dilemmas.

As an example, the fact that people will flee cities as the frequency and intensity of urban heat waves increase is expected; such migrations are foreseeable adaptation strategies. Predicting what will happen at the next step, when those people arrive at new locations, is a challenge to current planners. The fact is that it is possible to prepare for this influx of climate-related displaced persons into new regions so that they do not have to start their new lives unprepared or in poverty or subjected to deprivation or discrimination. The bottom line of a strategic adaptive strategy would be to plan to provide a "soft landing" to those who have no choice but to migrate in order to adapt to the impacts of global warming on society and on environment. Thus, a need exists *now* to identify future soft landings for those who are most vulnerable to changes in climate and the environment and who will be displaced from their normal activities, forced to adapt to the new conditions of wherever they are forced to move.

Another concern is how the various mitigation and adaptation measures that exist in the crop and livestock, forestry, fisheries, bioenergy and other areas outside the "agricultural" sectors, such as biodiversity, might affect the food security of vulnerable people, both positively (win-win situations) and negatively (trade-offs and conflict situations). In other words, how adaptation in one sector might affect the possibility of adaptation in other non-food related sectors must be considered. As noted earlier, governments will not have the resources to address all at once all of the potential impacts of climate change that are likely to affect their territories or citizens. They must, therefore, prioritize their responses. This necessity requires an "adaptation in parts" approach; that is, governments must choose to focus their assistance on the most at-risk segments of their populations and regions with respect to improving their overall food security. The challenge with "adaptation in parts" is identifying reliable indicators that define successes, weaknesses, opportunities and constraints caused by the interdependencies and controversies that exist among various sectors' adaptation and mitigation measures. Specifically, governments must draw on dependable tools to assess who is most at-risk, where they are, how to deliver services to them, etc. In this regard, a SWOC/T assessment would be very informative.

In line with the "Four Laws of Ecology," governments have to realize that whatever adaptive strategies or tactics they ultimately pursue, those strategies and tactics will, as noted earlier, generate their own impacts in other sectors. Mechanisms, processes or secondary assessments must be undertaken to identify second-order (ripple or downstream) impacts.

THE MARINE ENVIRONMENT AND GLOBAL WARMING: IMPLICATIONS

The ocean is a major repository (or sink) for carbon dioxide. Researchers are trying to determine how global warming will influence the behavior of ocean currents and air-sea interactions such as those associated with El Niño events. This knowledge will help us to understand the future impacts of global warming on the living marine resources contribution to food security.

An international symposium on "Effects of Climate Change on the World's Oceans, held in the summer of 2008 underlined that researchers only "have a rudimentary understanding of the sensitivity and adaptability of natural and managed ecosystems to climate change." It noted that:

> An assessment of the consequences of climate change on the World's Oceans has a high scientific and social relevance and is urgently needed. Although we are beginning to document the local effects and consequences of climate change on the functioning of marine ecosystems, there is no comprehensive vision at the global scale, and only limited ability to forecast the effects of climate change.
>
> To close this gap ... the symposium brought together results from observations, analyses and model simulations, at a global scale, and included discussion of the climate change scenarios and the possibilities for mitigating and protecting the marine environment and living marine resources.
> [http://www.pices.int/meetings/international_symposia/2008_symposia/ Climate_change/climate_publications.aspx]

In summer 2009, the World Ocean Conference and Coral Triangle Initiative (WOC-CTI) summit recognized the importance and interactions of "Climate Change Impacts to Oceans and The Role of Oceans to Climate Change". The overall goal of the World Ocean Conference (WOC) was to provide a forum for the international community to discuss current issues in the marine field which are related to climate change, and how the world can wisely utilize the ocean to weather crises. Furthermore, conference organizers expected to create more commitments from participating governments and institutions to work together to improve marine resource management. Inline with the expectations, representatives from 76 countries at the inaugural World Ocean Conference adopted the Manado Ocean Declaration (MOD) and reiterated the importance of achieving an effective outcome at the COP15 (Conference of Parties) of the UNFCCC in Copenhagen, and invited parties at COP15 to consider how the coastal and ocean dimensions could be appropriately reflected in their decisions. The Manado Ocean Declaration further recognized the role of ocean resources to enhance global food security and, concerned about environmental degradation and increased risks to global food security, the declaration read:

> *Recognizing that oceans and coasts provide valuable resources and services to support human populations, particularly coastal communities that depend heavily on them, and that the sustainable use of marine living resources will enhance global food security and contribute towards poverty reduction for present and future generations,..*
>
> *Equally concerned over marine ecosystems and living resources being affected by sea level rise, increased water temperature, ocean acidification, changing weather patterns, and other variations that may result from climate change, and how these alterations may aggravate the existing pressures of marine environmental degradation and increase risks to global food security, economic prosperity, and the well-being of human populations. [http://www.woc2009.org/MANADO_OCEAN_DECLARATION.pdf]*

In addition to actual international commitments by nations, notions such as "best practices" and "forecasting by analogy" can provide guidance at least in the near to midterm future fisheries management techniques at the local scale. Regarding food security, means must be found to reduce losses that result from both the discard of by-catches and the various fish capture and processing techniques that are known to be inefficient and wasteful. The oceans are an important component in the food security equation of many societies, as suggested by the discussion of "ghost acres" above.

IGNORANCE VS. "IGNORE-ANCE"

Many decision makers are ignorant about climate change, that is, they still do not really know much about climate change, despite all the media coverage about it over the past three decades. Of those who do know, their perceptions of the severity and urgency of the problem vary from strong believers to nominal believers in the ability of human societies to change the global climate. Even though they are going to have to make difficult and sometimes unpopular decisions, they lack the information and training they need to better understand and evaluate the science of the crisis as well as the potential severity of its impacts on food security, energy, and the food/energy nexus.

A lack of understanding of global warming is in many ways understandable, but such a lack can be readily overcome with additional knowledge transfer.

Ignore-ance, however, presents a very different problem. What ignore-ance means is that there are decision makers who understand the basic science of global warming and its projected consequences for society and for the ecosystems on which they depend, but they simply ignore it, caring more about re-election concerns or issues that are of immediate concern to their constituents who often do not themselves understand the gravity of the climate crisis. Some of these leaders may also believe that the impacts of climate change will not play out as a "worst case" scenario, and that societies will be able to keep up with incremental changes in the temperature and the environment. Often, the conflict such policy-makers face is between near-term benefits (their own) and longer-term costs (to the policy-makers who follow them).

©FAO/G. Napolitano

RAINFED FARMERS IN SUB-SAHARAN AFRICA

Climate change affects everyone. Hardest hit will be rainfed farmers who cover 96 percent of all cultivated land in sub-Saharan Africa, 87 percent in South America and 61 percent in Asia.

CHAPTER 5

SHORT-TERM AND LONG-TERM POLICY OPTIONS

ALL CLIMATE IMPACTS OF CONCERN TO POLICY-MAKERS ARE LOCAL

A common expression suggests "all politics are local"; a similar statement could be made for political concern about the impacts of global warming. As suggested earlier, global average statistics about global warming's influence on temperature and on the hydrologic cycle are not particularly useful for policy makers at the national to local levels. Thus, climate change concerns eventually center on what the impacts might be at various sub-national levels.

A search on the Internet using the phrase "What will be the manifestations of climate change?" yields many articles suggesting the possible impacts in different countries (and regions within countries) of global warming of a couple of degrees Fahrenheit. For example, one UK report notes the following:

> *The ways in which climate change manifests itself will vary dramatically from region to region in the UK, according to the experts. City centres will become hotter, changes in agricultural practices will alter the rural landscape beyond recognition and some coastal areas could find themselves completely submerged – a major concern for the economy, as half of Britain's prime agricultural land is below the five-metre contour.*

> *Many of Britain's large industrial plants – from oil refineries to nuclear power stations – are also concentrated on the coastline, and may have to be moved or defended in the face of rising sea levels. Some of the country's largest landfill sites are situated on former coastal marshes, which could spell environmental disaster in the event they become swamped.*

> *Droughts are expected to increase in the south, especially in the summer. The north and west are more likely to suffer from abundant and intense rainfall.*

The combination of sea level rise with high tides and changes in winds means severe flooding – the likes of which has plunged much of Britain under water in recent weeks – will become a regular threat. [http://news. uk.msn.com/How-climate-change-will-affect-regions-in-the-UK.aspx]

When considering how climate change is going to affect the UK, it's useful to understand the risks that current climate already poses to individuals, landscapes, organizations and the economy, before moving on to explore future climate risks. This report provides a valuable source of information to support this first step, stimulating better understanding of how the UK's climate affects our everyday lives (http://www.ukcip.org.uk/images/ stories/08_pdfs/Trends.pdf)

Korean government agencies are also identifying potential impacts of global warming that will affect their peninsula:

Q. How will global warming manifest itself on the Korean peninsula?
A: When the temperature rises, torrential rain becomes more frequent and the sea level rises. The temperature of the sea will go up. When the rise reaches 2-3 degrees, 20-30 percent of existing species will become extinct. The species of fish caught in the sea will change and a lot of jellyfish will appear. Fruit growing regions will also change, as seen in the way apples are now grown in Yanggu, Gangwon Province.

A developing country example was provided when, in December 2007, government and other officials met in Kinshasa, Democratic Republic of the Congo to discuss the possible impacts of global warming on activities in the DRC. The UNDP DRC director stated his concern about climate change in the following way: "What is climate change in the DRC? How does it manifest itself, and how can we, civil society, politicians and the international community, come together to plan and find a solution?" (http://allafrica. com/stories/printable/200712210975.html).

Yet another example of concern was provided in a PowerPoint format by a Caribbean climate change project, PRECIS-Caribbean (http://precis. insmet.cu/eng/Precis-Caribe.htm):

How will climate change manifest itself in the region?
Will a future Caribbean climate be:
•Hotter/Cooler?
•Wetter/Drier?
Important (we feel) if we are to address vulnerability and adaptation.

The point of using these disparate geographic examples in which the same question – "How will global warming (or climate change) manifest itself" in my region? – is posed to support the contention that locales worldwide are already engaging in self-appraisals about what they might expect with an even warmer atmosphere, having unknowingly been coping with a changing climate during the past few decades.

WORKING WITH CHANGE, NOT AGAINST IT

The phrase "climate change" raises eyebrows and interest now as never before. The word "change" is responsible for this awareness. Most people, institutions and governments fear change that they do not control, which should be remembered in discussions about climate change.

In the mid twentieth century, Eric Hoffer, an American migratory worker and self-taught social philosopher, wrote a book entitled *The Ordeal of Change* in which he discussed how people fear even the smallest changes to their routine or way of life (Hoffer, 1963). He wrote about the fear he faced as a migrant worker in California during severe multi-year droughts in the United States. He had finished picking peas on one farm and was about to move to pick beans next on a different farm, but he was afraid he would not be able to pick beans. Most people today might not see this shift in work as an insurmountable change, but it was to him. Today with a changing global climate, the fear is mounting in civil society and among its representatives of a new kind of unprecedented change that will have more serious implications for societies and their citizens. How will members of society feel when their lives are forced to change because of a warming climate?

Change can take place in many ways: it can be an abrupt, step-like change or a long, drawn-out affair. An abrupt change is clearly a crisis for a society. Some scientific reports warn of abrupt climate changes occurring in relatively short time periods (on a scale of decades), if one or another tipping point in the global climate regime is reached. Unfortunately, scientists and decision makers do not have adequate local to national reliable, detailed information about the possible impacts to take immediate actions to minimize potential damage. Again, this is yet another reason that resilient adaptation provides a level of flexibility necessary for an effective response to future climate uncertainties.

For slow-onset changes, different problems arise. First of all, they are preventable and reversible up to a level of degradation. Second, the consideration that incremental changes would eventually lead to major environmental changes, if not full-blown crises that demand concentrated attention and require a large amount of funds to address, seems difficult for policy-makers to accept. Third, and most important, is the fact that governments do not have a good track record of dealing with creeping, incremental but cumulative changes in the environment, at least not until those changes have reached the stage of environmental crisis. Rates of change in greenhouse gas emissions, in local temperature and precipitation, in ecosystem functioning, and in demographics are extremely important to monitor for identifying impacts and response strategies in a timely way and then prioritizing them.

The point is that climate-related change will not directly affect all people within a given region or country at the same time or in the same way. In the near future, policy-makers will have to convey this idea to local people and their leaders, but first researchers from various fields will have to determine effective ways to convey this notion to policy practitioners (Box 3).

APPROACHING ADAPTATION AND MITIGATION PLANNING WITH EYES WIDE OPEN

Adaptation to climate change has been a serious research topic for about three decades, even though widespread government interest was recently sharply elevated to new heights with the issuance of the IPCC 4th Assessment, the widespread viewing of Al Gore's *An Inconvenient Truth* (*Al Gore, 2006*), and the awarding of the Nobel Prize to both Gore and the IPCC process. With human-induced warming of the Earth's atmosphere, awareness and interest alone will not be enough, however, as known hazards are likely to take on continually new characteristics, becoming, for example, more intense, more frequent, and perhaps occurring in new locations within a country – or possibly, with some luck, even disappearing from a country or region altogether.

BOX 3

EDUCATION

Education refers to informing policy-makers about what they need to know about baseline conditions. This is viewed as a prior question that should be dealt with before making policy.

- National policy-makers need country-specific information, much of which is already available but may not be available to them in a central location and in readily usable form.
- Governments and individuals alike must remain alert to subtle changes in the environment and to the dynamics of the invisible boundary line at the human interface with climate-sensitive ecosystems.
- Governments must pay more attention to coping with slow-onset, low-grade (creeping) changes in climate, water and weather.
- Government officials would benefit from being reminded of *The Four Laws of Ecology* to warn them about the limits of tampering with natural processes.
- SWOC/T (Strengths, Weaknesses, Opportunities, and Constraints/Threats) assessments are useful tools for identifying and responding to a country's SWOC/T.
- Developing an initial set of strategies and tactics for coping with climate change impacts on food security is only the beginning step of an ongoing process.
- Existing "best practices" should be viewed as tools to provide tactical responses (short-term) to a changing environment.
- Policy-makers must reduce a society's fear of change because climate has always and will continue to change, and society has and must continue to adapt.
- Collaborative and strategic partnerships, domestic and international, can strengthen food security in face of an uncertain future (i.e. disaster-avoidance diplomacy).
- Beware of science-related "fads" for adaptation to or mitigation of global warming. Evaluate before you invest in new adaptation schemes.

Researchers and policy-makers concerned about global warming have focused on one or another aspect of its potential impacts, especially those concerning aspects of fisheries adaptation, urban adaptation, coastal adaptation, agricultural adaptation, health adaptation, public safety adaptation, aquaculture adaptation, tourism adaptation and arid lands adaptation, among many others. Adaptation discussions also focus on specific ecosystems, such as coral reefs, mangroves, tropical rainforests, Polar Regions, mountains, rivers, lakes and marine ecosystems, and so forth. For its part, UNDP's Global Environment Facility (GEF) has noted that its funding would focus on particularly vulnerable regions, sectors, geographic areas, ecosystems and communities.

A different approach to identifying and categorizing adaptation practices could involve placing specific information about adaptation into one or more of the following categories: Adaptation science, adaptation impacts, adaptation policy & law, adaptation politics, adaptation economics, adaptation technology, and adaptation ethics & equity. Yet another approach would be to list the known hazards to a country or region (super- or sub-national) and evaluate the effectiveness of the government's responses to occurrences in the recent past. Such a strategy is intimated in a Chinese proverb: "To know the road ahead, ask those coming back." This strategy would involve identifying the strengths and weaknesses of a society's response mechanisms, and would be a positive action to identify constraints and weaknesses, which could, in turn, help to identify strengths and opportunities that exist or could be developed to overcome them. In addition to a country's recent history of coping with its climate-related and other hazards there is also the possibility of learning about successful mitigation and adaptation policies and practices from other regions and countries that have had to face similar climate-, water- or weather-related hazards.

Clearly, many approaches are possible in developing reliable and effective action plans for mitigation and adaptation to a changing climate. The point is that, though some may prove more effective than others, governments can pro-actively pursue any of these different approaches to protect their societies and economies against these changes' impacts. Because each approach will likely have significant overlap with others, governments will need to determine how best to identify, develop and pursue adaptive strategies and tactics that are best suited to their specific situations and needs.

The 3rd Law of Ecology, that "Nature knows best," forces decision makers to take a closer look at the recent history of the interactions of food security, food insecurity and climate (variability, fluctuations, extremes and change). Many examples from history can be cited where people believed that they were able to dominate nature and grow what they wanted wherever they chose to grow it with little regard for the long-term stability of the environments they chose to exploit, only to discover the reality of humanity's vulnerability to the whims of the environment. The Soviet Union's "Virgin Lands Scheme" of the 1960s and 1970s is an example of a failed attempt to dominate nature. The diversion of the two rivers that feed the Aral Sea in Central Asia, thereby depriving the sea of its water supply, is another example of a failed attempt to dominate nature. The river diversions have led to the near-disappearance of what was once the world's fourth largest inland sea and to a grave deprivation, both in terms of health and livelihoods, for those who once relied upon it for their sound existence. Yet another example is mangrove destruction in order to create shrimp farms in areas around the world. Unfortunately, many similar examples could be cited because individuals and societies have for too long not respected the limits to which nature can be transformed without harming it.

ADAPTATION AND MITIGATION STRATEGIES AS OUTPUTS

Adaptation and mitigation strategies as outputs refers to the specific reports, conferences and workshops designed to produce policies that are to be pursued to minimize, if not avert, the adverse impacts of global warming on society and on ecosystems. One challenge is to identify suitable indicators to define success and/or limitations of adaptation and mitigation caused by the interdependencies as well as controversies and conflicts that either presently exist or might arise with other sectors' responsibilities under global warming scenarios.

The first concern of a government should be the safety and well-being of its citizens; the second should be its territorial integrity and protection of infrastructure. Policy-makers have often relied on experts within and outside of government to help identify effective strategic responses to climate variability and extremes in executing these primary and secondary mandates, and now they must do the same for climate change. Reports and other written resources describing these plans are outputs, which are readily quantified; whether the actions outlined in these outputs will prove effective in practice must be tested by both nature and human activities.

The Australian government has provided a brief set of actions that should be considered when establishing adaptive policies for climate change (Australian Greenhouse Office, 2007). The steps in adaptation are as follows: plan early, be systematic and strategic, use the best information, and be flexible. (http://www.greenhouse.gov.au /impacts/howtoadapt/index.html). Similarly, the Danish government has identified a sampling of possible adaptation options that it sees as linked to poverty reduction and economic development:

- Protect against sea-level rise, including salt-water intrusion into water supplies.
- Strengthen primary health care in preparation for the potential spread of vector-borne diseases. Change building codes to withstand extreme weather events, rehabilitation of natural ecosystem such as mangroves to reduce the impacts of storm surges. Redesign infrastructure in regions expected to become wetter with climate change.
- Develop new crops, cropping strategies and insurance schemes.
- Manage water resources for sustainable supplies.

The Danish government believes these adaptations to climate change will be effective and are necessary to manage climate, water and weather-related risk.

Outputs relate to the measure of success of a decision making process. Agreements can be reached, laws can be made, and Plans of Action can be agreed upon; indeed, these are the instruments through which the success of outputs are typically measured. The successes of the decision making process to some extent depend on the motivation, ways and means through which climate change adaptation and mitigation strategies are provided to the policy makers. A priority list of "illustrations" for policy makers relates to adaptation and mitigation of climate change impacts on food security is given in the Box 4.

ADAPTATION AND MITIGATION STRATEGIES AS OUTCOMES

Outcomes measure the effectiveness of the outputs of a decision making process. Laws can be passed by legislatures, for instance, but if no provisions provide for enforcement of or compliance with those laws, the likelihood of favorable outcomes is diminished. Societies hope to cope effectively with extreme climate, water and weather related events, which means that they

BOX 4
ILLUSTRATIONS

Illustrations refer to the ways that information can be exposed to policymakers. Examples of successful adaptations to change that have taken place can be into this category. It is also where successful mitigation activities that have been put carried out can be highlighted. It is also a place for example of mistakes in the incorrect or non-use of climate-related information.

- "Mapping" resiliency in a country in addition to "mapping" vulnerability is useful and important; they are not in opposition to one another.
- Rates of change need to be monitored because they are as important as processes of change when it comes to strategic planning for adaptation and mitigation.
- The future environmental impacts of new land uses can actually be identified, since those impacts have likely occurred elsewhere on the globe. Research focused on finding those places should be supported.
- A SWOC/T assessment can also be valuable to minimize the chance of mal-adaptation.
- Invest in ways to improve scientist-to-policymaker communications to improve understanding of scientific findings related to food security.
- List known hazards to a country or a region and evaluate the level of effectiveness of previous governmental (and societal) responses to recent extreme events.
- Identify suitable indicators to define successes and limitations of adaptation and mitigation caused by interdependencies and controversies/conflicts that either presently exist or might arise through other sectors' responsibilities under global warming.
- Tradeoffs must be made explicit to policymakers for proposed adaptation and mitigation strategies and tactics.
- The FAO can provide support for undertaking national baseline assessments for greenhouse gas emissions from agriculture-related activities.

BOX 5

RECOMMENDATIONS

Recommendations refer to actual suggestions for policy-driven strategic thinking and action for mitigation and adaptation to climate change related to food security and biofuels.

■ Policymakers must introduce and foster the notion of "resilient adaptation" throughout their ministries and agencies as well as in civil society.

■ Policymakers must require researchers as well as their own agencies to identify and focus on the protection of Areas of Concern (AOCs) to stop an environmental degradation from becoming a hotspot that will adversely affect food security, forestry and fisheries.

■ Explicitly acknowledge agriculture-related controversies and conflicts and then put them into the context of global warming to generate cooperation. Otherwise, proponents will continue see themselves as locked in an apparent zero-sum game with one side winning at the expense of all others.

■ Refrain from identifying winners and losers of climate change until objective measures of what it means to win or lose have been identified.

■ Keep in mind the Precautionary Principle. In other words, do not use scientific uncertainty as an excuse to avoid decision making when using climate change scenarios for making strategic planning.

■ Given the limited funds available, most governments will need to prioritize their adaptive capacity building to climate change impacts and then undertake "adaptation in parts."

■ Additional assessments are needed to discover second-order (downstream) impacts of adaptation and mitigation strategies and tactics.

■ Call for and support an assessment that identifies both the obvious and the hidden reasons why hazard-and disaster-related lessons are identified after each disaster but are often not applied (used).

- Require that all new projects affecting the environment, including forestry and fisheries, include a "Food Security Impact Assessment" (FSIA).
- Policymakers must be made aware of the importance of changes in seasonality and must consider this a high priority concern about climate change, since people and economies are align with the expected natural flow of the seasons.
- Consider prevention along with mitigation and adaptation because new activities that are known to produce greenhouse gas emissions can be blocked.
- Policymakers must harmonize the activities of their ministries, agencies and bureaucratic units with the rules used to govern administrative jurisdictions.

want to emerge after an extreme event in the same or better condition as they were in before the onset of the extreme event. In the absence of perfect information in the form of a reliable early warning, for example, costly adverse impacts can be expected. Outcomes are more difficult to measure than outputs, though, interestingly, they are a direct consequence of the level of effectiveness and efficiency of adaptive strategies and tactics.

Issues surrounding adaptive strategies are related both to reducing vulnerability and to increasing societal resilience; importantly, however, reducing vulnerability does not necessarily increase resilience, and increasing resilience does not necessarily reduce vulnerability. In other words, these concepts are not opposite sides of the same coin. The optimal outcome of an effective strategic response to climate change would be a lowered level of vulnerability in regard to food security and a higher level of resilience in the face of climate change's potential impacts. Perhaps "resilient adaptation," a notion borrowed from psychotherapy, defines the optimal outcome. A set of comments on recommendations for policy-makers (Box 5) relates to climate change adaptation and mitigation. The comments on recommendations are not mutually exclusive with previous categories of education and illustration.

©FAO/A. Mihich

A TRADITIONAL BAMBOO WATER MILL IN VIET NAM

Traditional strategies had evolved long before recorded history for coping both with climate variability as well as climate extremes.

LESSONS LEARNED ABOUT "LESSONS LEARNED":

Lessons identified from climate, water and weather impacts are NOT to be considered as lessons learned. They are only lessons that have been identified for attention, until they have been addressed. Just about every hazard or disaster related assessment ends with a list of lessons learned and recommendations for the future (Glantz, 2000); however, the phrase "lessons learned," has become part of the problem of addressing issues related to early warnings and coping with future hazards.

The phrase "lessons learned" suggests that someone (we do not know exactly who this person is) is ensuring that a problem identified from a past experience will not happen again (at least in terms of severity of impacts). Once lessons have been identified and publicly broadcast, however, who is expected to listen and take action to address those lessons? Furthermore, are those who identified those lessons in a position to influence those who are in a position to implement change? The foreseeable truth is that no one, in the end, may have been delegated that responsibility, even after a lesson has been clearly identified. As time passes, interest in that particular disaster event as well as concern about its victims become overshadowed by more pressing issues and newly emergent hazards.

A preliminary review of case-specific lessons learned from a sixteen-country El Nino-related impacts assessment exposes the reality that many of those often costly lessons learned were really not learned in the true sense of the term. A multi-decade review that looks at previous disasters either in the same location or elsewhere would most likely uncover similar previously identified lessons. The point is that while some lessons are identified and applied – that is, truly learned – as a result of a given assessment, many of the directives derived from these costly lessons go unfunded or under-funded, which means that they are unused and eventually forgotten. They are rediscovered during reviews following the next similar disaster for which those lessons had already been identified. And so the cycle continues.

An assessment is urgently needed that focuses on identifying both the obvious and the hidden reasons why hazard- and disaster-related lessons are so often identified but so seldom learned, and end up gathering dust on library shelves.

ADAPTING TO AND MITIGATION OF CLIMATE CHANGE: "WHAT OUGHT TO BE" VERSUS "WHAT IS"

On paper, problems in the world are easy to solve. Planning activities to address them are quite thorough, often almost perfect. Numerous details are identified that must be attended to, and paths toward achieving the objectives laid out in those details are well-defined. That is what happens on paper. The impediment to this scenario is, of course, that the world does not exist only on paper, and problems in the real world often – always, really – arise when it comes to implementing the various aspects of these perfect paper plans. Plans (like scenarios) can be viewed as "what ought to take place" usually in a perfect world setting.

The reality is that in most cases these plans cannot be implemented as proposed, given the numerous potential constraints and hurdles – economic, political, social, cultural, infrastructural and bureaucratic – that must be overcome. Many examples can be cited of how the best laid plans for dealing with the impacts of a climate, water or weather disaster (drought-related food insecurity, hurricane impacts, the collapse of a fishery, an infectious disease outbreak) proved extraordinarily difficult to carry out effectively. The same problem applies to early warning systems: they work effectively if communication is timely and the targeting of at-risk populations is efficient, but neither effectiveness nor efficiency has ever been achieved at a perfect level in so imperfect a reality as the one in which these problems arise.

In theory, designing ways to enhance food security is also easy, but in practice many real constraints must be overcome to achieve the desired outcomes as described on paper. Looking at recent situations and the way they were responded to, and comparing them to a perfect response in the absence of constraints, provides an opportunity to identify the hidden bottlenecks that hinder the achievement of food security in the face of a variable and changing climate (Box 6).

BOX 6

RAMIFICATIONS

Ramifications refer to aspects of policy that, if neglected, would yield negative feedback to policymakers. The question that policymakers must ask is "what are the consequences of not doing this suggestion?"

- Every country needs to prioritize its hazards according to its own criteria, such as in terms of likelihood of occurrence and severity of impacts on citizens, infrastructures and ecosystems.
- The overriding objective for focusing on food security related to hotspots is to avoid creating new hotspots where they do not yet exist.
- Policymakers must not panic as they prepare for changes in the near and long term.
- Adaptation to change has to be appropriate to specific hazards or threats for a given period of time.
- There has been no attempt to systematically identify, region by region, which climate changes might be advantageous and which might be harmful.
- Decision makers must maintain a degree of flexibility in the implementation of their adaptation and mitigation strategies and tactics.
- Consideration must be made of how adaptation in one sector might affect the possibility for effective adaptation in another sector.
- Resist the pressure on decision makers to go for short-term benefits at the expense of long-term costs.
- Foster a cross-sectoral (multidisciplinary) approach that matches cross-sectoral aspects of and need for adaptation and mitigation. Such an approach will foster a broader, more appropriate approach to adaptive capacity building.
- Assure awareness and dissemination of conference and workshop proceedings about adaptation and mitigation to produce understanding and to reduce fear among both the general public and professionals.
- Policymakers must not only enhance agriculture's mitigation role but must also reduce the vulnerability of poor and marginalized people to food insecurity.

WHY SOME SOLUTIONS TO ACHIEVING FOOD SECURITY ARE KNOWN BUT NOT APPLIED

The problems and prospects for a food-secure country can be made explicit by looking at three levels of analysis: individuals & groups, governments & agencies, and the international community. Into these categories, which are not mutually exclusive, might be added, at least at the individual level, human nature, such as, for example, the desire or need for short-term exploitation of land and water that overrides long term concerns about sustainability. In such a case, government leaders might choose to use fertile lands or deforest forested areas to produce cash crops such as biofuels or flowers; at the government level, bureaucratic units focus on their areas of concern with less regard for the impacts on the areas of concern off other bureaucratic units; at the international community level, food and feed are provided as trade and aid, the amounts of both being dependent on variable climate and market price factors. By looking at the food security and climate change issue according to these three levels of social organization, hidden obstacles to effective policy-making can be identified.

Effective national policy-making instruments and institutional arrangements are needed to override sole dependence on a sector by sector approach to enable the identification of crosscutting, downstream issues and impacts, including effective ways to address cross-sectoral issues, and long range time horizons that can strengthen local decision making mechanisms to ensure effective and rapid responses on the ground. A cross-cutting multi-sectoral approach can also serve to strengthen sectors and to foster a broader multidisciplinary perspective (FAO issues paper).

Most observers of the global food situation believe that enough food is produced around the globe to feed every person adequately, but transporting food and feed at low cost from surplus regions to food-deficit regions is expensive and is done usually only in times of an impending famine, not for situations categorized as chronic hunger.

One prior question (a question that precedes action) about coping with the impacts on food security related to climate change has become clear: Policy-makers must decide how they intend to organize in order to plan strategic responses. Are they going to rely on existing traditional institutional arrangements, such as their existing sector-based governmental bureaucracies? Or are they going to seek new institutional arrangements to develop strategic and tactical plans for climate-, water- and weather-related global changes.

Another important point is that even if all human-induced greenhouse gas emissions could be stopped from entering the atmosphere today, the global climate will still continue to warm for much of the rest of the twenty-first century, because of the endurance of the various GHGs in the atmosphere. Again, there is no proverbial 'silver bullet' solution for controlling climate change or for coping with its impacts. Many solutions are yet to be identified.

KEY TAKE-HOME MESSAGES FROM THE FAO HIGH LEVEL CONFERENCE

In order to put agriculture, forestry, fisheries and food security on the international climate change agenda, the Food and Agriculture Organization of the United Nations (FAO), in cooperation with the Consultative Group on International Agricultural Research (CGIAR), the International Fund for Agricultural Development (IFAD) and the World Food Programme (WFP), organized a High-Level Conference on "World Food Security: The Challenges of Climate Change and Bioenergy" held at FAO Headquarters in Rome, Italy (June 2008). The conference brought together world leaders, policy makers and experts from many disciplines and discussed the challenges that climate change, bioenergy and soaring food prices pose to world food security. The major outcomes of the conference are: (i) Identification of the new challenges facing world food security, (ii) A better understanding of the nexus between food security, climate change and bioenergy, (iii) Discussion of required policies, strategies and programmes for ensuring world food security, in particular measures to address soaring food prices and (iv) A declaration on "World Food Security and required actions."

The following list highlights some of the summary of typology of management and policy options relevant to country level actions synthesized based on the discussions during the FAO's expert meetings that preceded the high level conference:

Climate change adaptation and mitigation

- Adaptation measures need to focus on
  - climate change "hot spots" analysis,
  - early warning systems,
  - disaster risk management,
  - rural investments: crop insurance, incentives to adopt better agricultural and land use practices,
  - building capacity and awareness on climate change adaptation,
  - extension and research services at national level - data collection, monitoring, analysis and dissemination,
  - Soil Carbon Sequestration – potential option for mitigation in agriculture.

Climate change, water and food security

- Integration of adaptation and mitigation measures for agricultural water, management in national development plans,
- Technical and management measures to improve the water use efficiency in rainfed and irrigated agriculture,
- Knowledge on climate change and water, and share good practices among countries and regions,
- Risk management in national policies through better monitoring networks,
- Adaptation funds to meet the challenges of water and food security under climate change.

Climate change and disaster risk management

- Better understanding of climate change impacts at local level,
- Diversifying livelihoods and adapting agricultural, fishing and forestry practices,
- Improving and expanding weather and climate forecasting and early warning systems,
- Contingency plans and disaster risk management plans in agriculture taking into consideration new and evolving risk scenarios,
- Adjustment of land use plans,
- Cost/benefit analysis on structural mitigation measures.

Climate-related transboundary pests and diseases:
- Strengthening national animal and plant health services,
- Focusing on basic sciences - taxonomy, modelling, population ecology and epidemiology,
- Consolidating and organizing national animal and plant health services,
- Investment in early control and detection systems, including broader inspections,

Climate change, fisheries and aquaculture:
- Adaptation strategies based on ecosystem approach,
- Understanding and anticipating ecological change and developing appropriate management responses.

Climate change, biofuels and land,
- Sound land tenure policies and planning,
- Greater land tenure security to mitigate climate change,
- Enabling and encouraging investments in sustainable land use practices.

Bioenergy and food security
- Ensuring that bioenergy is developed sustainably,
- Safeguarding food security and ensuring that benefits include market and technology promotion and encouraging participatory processes.

Climate change and biodiversity
- Assessment of distribution of biodiversity for food and agriculture both in the wild and in the fields,
- Assess its vulnerability to climate change,
- Biodiversity distribution mapping with different climate change scenarios.

Following significant discussion and negotiations, the conference concluded with the adoption by acclamation of a declaration calling on the international community to increase assistance for developing countries, in particular the least developed countries and those that are most negatively affected by high food prices. The declaration reads:

COPING WITH A CHANGING CLIMATE: CONSIDERATIONS FOR ADAPTATION AND MITIGATION IN AGRICULTURE

"There is an urgent need to help developing countries and countries in transition expand agriculture and food production, and to increase investment in agriculture, agribusiness and rural development, from both public and private sources," according to the declaration and noted that "It is essential to address the fundamental question of how to increase the resilience of present food production systems to challenges posed by climate change."

On climate change, the conference urged governments to assign appropriate priority to the agriculture, forestry and fisheries sectors, in order to create opportunities for the world's smallholder farmers and fishers, including indigenous people, in particular vulnerable areas, to participate in, and benefit from financial mechanisms and investment flows to support climate change adaptation, mitigation and technology development, transfer and dissemination.

On the issue of biofuels, the conference concluded that it is essential to address the challenges and opportunities posed by biofuels, in view of the world's food security, energy and sustainable development needs. The members are convinced that in-depth studies are necessary to ensure that production and use of biofuels is sustainable and take into account the need to achieve and maintain global food security. The conference called upon relevant inter-governmental organizations, national governments, partnerships, the private sector, and civil society, to foster a coherent, effective and results-oriented international dialogue on biofuels in the context of food security and sustainable development needs.

A "REALITY CHECK"

"Nature can look after the needs of people. It cannot look after the greeds of people." (Gandhi)

Some topics are more or less treated as "taboo" when discussing food security issues, such as corruption, politics, the notion of "the greatest good for the greatest number," hidden subsidies, ethnic rivalries, greed, and population as an environmental issue. Numerous examples can be cited of how each of these topics has affected food security and the environment in countries around the world: During the Sahelian drought in the early 1970s, some West African countries continued to export cash crops while their citizens were perishing from starvation in a famine that resulted from severe, drought-related food shortages. As another example, large swaths of the Brazilian rainforest have been deforested by people fleeing drought in the Brazilian *Nordeste* in order to grow crops on lands that have soils known to be fragile for sustainable cultivation. In Indonesia, rainforest has been "torched" on purpose by then-corrupt government officials in order to develop illegal oil palm plantations on previously protected land, and in the United States, land has been taken out of production (put into a 'land bank') in order to control the price of grain in the marketplace. Finally, productive land has been converted to golf courses in Japan. And the list goes on and on.

Because every government has financial constraints, each has to make decisions involving trade-offs between costs and benefits. In tradeoffs between providing food security for 100 percent of a population and attempting to increase foreign exchange earnings, for example, the latter usually takes precedence because exchange earnings have been valued more highly than full nutritional capacity. Other trade-offs emerge, as questions policy-makers must negotiate answers to: Are city dwellers favored over rural inhabitants? Should Mangroves receive protection over shrimp farms? Is wheat a priority over teff, sorghum or millet? How are questions such as these to be answered?

An issue that generates much discussion but is nonetheless still considered taboo is population. The fact is that climate change will impact the availability of and access to adequate food and nutrition for most if not all people on the globe. Given increasing population numbers coupled with economic development prospects and growing demands of the affluent for food and fuel, food insecurity can be expected to increase under a "business as usual" scenario. Effective adaptation strategies can alleviate food insecurity *if* the "will" to do so exists at the international community level. "Ways" to cope effectively are either known or are likely to be soon developed, as food security becomes increasingly threatened by global warming. Population management has also been proposed by China as a "carbon sink" for which it could seek carbon credits. It noted that its one child per family policy has reduced the country's carbon emissions by the amount that the unborn populations would have produced over the course of their lifetimes. Many people disagree with such an approach in the name of protecting the environment.

In the end, policy-makers everywhere have always had to cope with many different and urgent issues involving competing interests for which they have to make decisions, often with incomplete information. How to mitigate and adapt to global warming is the latest such issue with which they have to deal.

INDIGENOUS PEOPLE IN ANDEAN MOUNTAINS

Indigenous peoples are among the first to suffer from increasingly harsh and erratic weather conditions, and a generalized lack of empowerment to claim goods and services to which other population groups have greater access.

6 A CONCLUDING THOUGHT:
NO ADAPTATION RECOMMENDATIONS WITHOUT RAMIFICATIONS

Every assessment of a natural or human-caused disaster begins and ends with a list of recommendations or lessons learned. The recommendations or lessons are about "how to get it right the next time there is a similar risk of a hazard becoming a disaster?" That is always the hope. That is always the dream.

Many of those recommendations or lessons learned are right on target in terms of requirements needed to reduce the adverse impacts of the hazards of concern. They are the result of serious scrutiny of hazards, their impacts and societal responses to them. They are the findings through serious discussion, brainstorming and plain common sense of what went right, what went wrong, and what wasn't considered (but should have been). For Katrina, for example, America's most costly and most embarrassing so-called natural disaster, one can find thousands of lessons learned from various levels of government, industries and businesses from local to global. That is the good news. However, it is, all too often, the good news in "theory" only, because most recommendations are not acted upon. Phrased a different way, the disaster lessons we have been calling 'lessons learned" are really not learned but only identified. When they are addressed they can legitimately be called lessons learned. Otherwise, they should be called "lessons identified".

The problem in all this is that when recommendations and lessons have been identified, many observers in all walks of life tend to think that the recommendations and lessons are being enacted in order to avoid similar hazard-related disasters in the future. The reality of the issue-attention cycle of any public and most policy makers lasts only a couple of years (as identified by Anthony Downs (noted scholar in public policy and public administration, and Fellow at the Brookings Institution in Washington D.C) in the early 1970s). The public tends to refocus its attention on other pressing issues, no longer focusing on the previous disaster or its recommendations. How then can we get policy makers to take recommendations and lessons more seriously? How can we get them to realize that not following up on the lessons can have costly consequences?

The vicious cycle is one of "disaster---lessons & recommendations---disaster---same lessons, ad infinitum. The recommendations and lessons remain but the political leaders change. Many of the same lessons appear decade after decade. Our children and our children's children will be reading the same sets of disaster-related recommendations and lessons that our predecessors and we have been identifying for decades. Enacting the following recommendation can help to end the cycle:

> *Recommendations (and lessons learned) should not be presented without comment on what the consequences might be, if the recommendations (and lessons) are not addressed. This way, decision makers can explicitly be made aware that there is also a likely cost for inaction when the next natural hazard turns into a national disaster. Succinctly stated, "RECOMMENDATIONS SHOULD NOT BE OFFERED WITHOUT NOTING THEIR RAMIFICATIONS, if recommendations are not implemented".*

The ramification, if the recommendation above is not acted upon, will be "business as usual" (BAU) with regard to identifying lessons and making recommendations in post-disaster assessments. This in turn means that policy makers in the future will likely continue to receive lists of lessons that had already been identified over previous decades and their societies will likely continue to remain at risk to the impacts of hazards for which risks could have been reduced, had those recommendations been pursued and the identified lessons applied.

©FAO/J. Isaac

FOREST-DWELLERS IN SOUTH EAST ASIA
Climate change impacts the livelihoods of forest-dwellers and adaptation of forest-dwellers and forest-dependent communities to climate change is a challenge.

REFERENCES

AlphaThink Consulting, 2003. High Performance Creativity and Decision Making. Website accessed 1/2008: http://alphathink.com/Frame-944278-servicespage944278.html? refresh=1193338038490

Australian Greenhouse Office, 2007. How to adapt. Website accessed 1/2008 at http://www.greenhouse.gov.au/impacts/howtoadapt/ Canberra, Australia: Department of the Environment and Water Resources.

Burns, D. 1989. Preface. *Climate and Food Security.* Papers presented at International Symposium on Climate Variability and Food Security in Developing Countries, 5-7 February 1987 in New Delhi, India. Manila: International Rice Research Institute. 602 pp.

Commoner, B. 1971. *The Closing Circle: Nature, Man, and Technology.* New York: A. Knopf. 326 pp.

DANIDA. 2005. Danish Climate and Development Action Programme: A Tool Kit for Climate Proofing, Danish Development Cooperation, Ministry of Foreign Affairs of Denmark, August 2005, 50 p. [http://www.netpublikationer.dk/um/5736/pdf/samlet.pdf]

FAO. 2003. Food security: Concepts and measurement. Chapter 2 in *Trade Reforms and Food Security: Conceptualizing the Linkages*, 25–33. Rome: Food and Agricultural Organization of the United Nations.

FAO. 2008. *Food, Energy and Climate: A New Equation,* FAO at work 2007 – 2008, Food and Agriculture Organization of the United Nations, Rome. 18pp.

Gallopin, G.C. 2006. Linkages between vulnerability, resilience, and adaptive capacity. *Global Environmental Change*, **16**, 293–303.

Glantz, M.H. (ed.) 1987. *Drought and Hunger in Africa: Denying famine a Future,* Cambridge, UK p.48.

Glantz, M.H. (ed.) 2000. *Once Burned, Twice Shy? Lessons Learned about the 1997–98 El Niño.* Tokyo, Japan: United Nations University. 294 pp.

Glantz, M.H. (ed.) 1990. *On Assessing Winners and Losers in the Context of Global Warming.* Report of workshop held June 1990 in St. Julians, Malta, p. 11. Boulder, CO: Environmental & Societal Impacts Group. 44 pp.

Glantz, M. H. 2003. *Guidelines for establishing audits of agricultural – environmental hotspots.* Food and Agriculture Organization of the United Nations, Rome, 28 pp.

Glantz, M.H. & Cullen, H. 2003. Zimbabwe's Food Crisis. *Environment*, 45, 1, 9-11.

Gore, A.A. 2006. An inconvenient truth. Film documentary about global warming. Distributed by Paramount Classics, 94 minutes.

Hoffer, E. 1963. *The Ordeal of Change*. Cutchogue, New York: Buccaneer Books. 117 pp.

IPCC. 2007a. Climate Change 2007: The Physical Science Basis. Contribution of Working Group I to the Fourth Assessment Report of the IPCC. *In* S. Solomon, D. Qin, M. Manning, Z. Chen, M. Marquis, K.B. Averyt, M. Tignor and H.L. Miller, eds. Cambridge, UK: Cambridge University Press. 996 pp.

IPCC. 2007b. Climate Change 2007: Synthesis Report. Contribution of Working Groups I, II and III to the Fourth Assessment Report of the Intergovernmental Panel on Climate Change. [Core Writing Team, Pachauri, R.K and Reisinger, A. (eds.)]. IPCC, Geneva, Switzerland, 104 pp.

IPCC. 2007c. Climate Change 2007: Impacts, Adaptation and Vulnerability. Contribution of Working Group II to the Fourth Assessment Report of the IPCC. *In* M.L. Parry, O.F. Canziani, J.P. Palutikof, P.J. van der Linden and C.E. Hanson, eds. Cambridge University Press, Cambridge, UK, 976pp.

IPCC. 2007d. Summary for Policy-makers, Climate Change 2007: Mitigation. Contribution of Working Group III to the Fourth Assessment Report of the IPCC. *In* B. Metz, O.R. Davidson, P.R. Bosch, R. dave, L.A. Meyer eds. Cambridge University Press, Cambridge, United Kingdom and New York, NY, USA.

IPCC. 2001. Climate Change 2001: Impacts, Adaptation & Vulnerability: Contribution of Working Group II to the Third Assessment Report of the IPCC. *In* J. J. McCarthy, O. F. Canziani, N. A. Leary, D. J. Dokken and K. S. White, eds. Cambridge, UK: Cambridge University Press. 1000 pp.

Lang, T & Heasman, M. 2004. *Food Wars. The global battle for mouths, mids and markets*. London: Earthscan, 2004. 364 p.

Lindblom, C. 1979. Still muddling through. *Public Administration Review*, 39(6), 517–525.

Lindblom, C.E. & Cohen, D.K. 1979. *Usable Knowledge: Social Science and Problem Solving*, New Haven: Yale University Press, passim.

Luthar, S. 2003. *Resilience and Vulnerability*, Cambridge, UK: Cambridge University Press. Passim.

MA (Millennium Ecosystem Assessment). 2005. What is the Millennium Ecosystem Assessment? Website accessed 1/2008. www.millenniumassessment.org/en/About.aspx

Myers, N. & Kent, J. 1995. *Environmental Exodus: An Emergent Crisis in the Global Arena*. Washington, DC:, The Climate Institute, p. 358. UN (United Nations), 1975: *Report of the World Food Conference*, held in Rome 5–16 November 1974. New York: United Nations.

Pielke Jr., R.A. 2003. "What is Climate Change? Policy Consequences of Differing Political and Scientific Definitions," 21 October. Center for Science Technology Policy, University of Colorado-Boulder/CIRES. www.climateadaptation.net/docs/papers/pielke.pdf

Petroski, H. 1992. *To engineer is human: the role of failure in successful design*, Vintage Books, New York, 1992.

UN Habitat, 2007. The Best Practices List. http://www.bestpractices.org

Wells, H.G. 1904. *The Food of the Gods and How It Came to Earth*. Chapter 1, accessed from http:// etext.library.adelaide.edu.au/w/wells/hg/food/marc.bib. Licensed under eBooks@Adelaide, 2006.

WLVC. 2003. World Lake Vision: A call to action, World Lake Vision Project Secretariat, World Lake Vision Committee, International Lake Environment Committee Foundation (ILEC), Kusatsu, Shiga, Japan, 36p. [http://www.ilec.or.jp/eg/wlv/complete/wlv_c_english.PDF]

World Bank, 2008. Climate change and development. From website accessed 1/2008. http://www.worldbank.org/

WRI (World Resources Institute). 2007. Agriculture and climate change: Greenhouse gas mitigation opportunities and the 2007 farm bill. WRI Policy Note: Climate and Agriculture, No.2, March 2007 [http://pdf.wri.org/agricultureandghgmitigation.pdf]

SMALL-SCALE FISHERMAN IN CAPE VERDE

Small scale fishermen living in coastal areas of small island developing states are threatened by raising sea levels, coral bleaching and salt water intrusion.

ANNEX

CLIMATE CHANGE AND FOOD SECURITY

It is generally accepted that climate change is the result of human activity including industrial output, car exhaust, and deforestation. These types of activities increase the concentrations of carbon dioxide, methane, nitrous oxide and other greenhouse gases (GHGs) in the atmosphere (IPCC, 2001). If the current trend in carbon emissions continues, temperatures will rise by about 1° C by the year 2030 and by 2° C by the next century. This increase, however, will probably have different impacts in different regions. Agricultural impacts, for example, will be more adverse in tropical areas than in temperate areas. Developed countries will largely benefit since cereal productivity is projected to rise in Canada, northern Europe and parts of Russia. In contrast, many of today's poorest developing countries are likely to be negatively affected in the next 50 – 100 years, with a reduction in the extent and potential productivity of cropland. Most severely affected will be sub-Saharan Africa due to its inability to adequately adapt through necessary resources or through greater food imports.

Problems facing farmers can be better understood if one considers the impact of climate change on weather or water. Precipitation, temperature and sunlight are the main factors behind agricultural production. Climate change can alter these factors causing essential threats to water availability, reduced agricultural productivity, spread of vector borne diseases to new areas, and increased flooding from sea level rise and even heavier rainfall. Climate variability is already the major cause of year-to-year fluctuations in production in both developed and developing countries. The largest reduction in cereal production will occur in developing countries, averaging about 10 percent, according to an FAO study (1996). A projected 2 – 3 percent reduction in African cereal production for 2020 is enough to put 10 million people at risk. These impacts would require adaptation efforts that in many cases will be hardly affordable for people living with little access to the necessary resources or savings. In fact, the real impact will be in areas where food production is already often marginal.

Some of the impacts of climate change on food production, which are already visible and seem to be advancing at a higher rate than previously anticipated include:

- Regional temperature rises at high northern latitudes and in the center of some continents;
- Increased heat stress to crop and livestock; e.g. higher night-time temperatures, which could adversely affect grain formation and other aspects of crop development;
- Possible decline in precipitation in some food-insecure areas such as southern Africa and the northern region of Latin America;
- Increased evapo-transpiration rates caused by higher temperatures, and lower soil moisture levels;
- Concentration of rainfall into a smaller number of rainy events with increases in the number of days with heavy rain, increasing erosion and flood risks;
- Changes in seasonal distribution of rainfall, with less falling in the main crop growing season;
- Sea level rise, aggravated by subsidence in parts of some densely populated flood-prone countries, displacing millions;
- Food production and supply disruption through more frequent and severe extreme events.

FOOD INSECURITY

Since food insecurity depends more on socio-economic conditions rather than on agroclimatic ones, the ways in which climate change can affect people's access to adequate food is rather complex. Future food security will mainly depend on the interrelationships between political and socioeconomic stability, technological progress, agricultural policies and prices, growth of per capita and national incomes, poverty reduction, women's education, trade and climate variability._Climate change, however, may affect the physical availability of food production by shifts in temperature and rainfall; people's access to food by lowering their incomes from coastal fishing because of rising sea levels; or lowering a country's foreign exchange earnings by the destruction of its export crops because of the rising frequency and intensity of tropical cyclones.

Some groups are particularly vulnerable to climate change: low-income groups in drought-prone areas with poor infrastructure and market distribution systems; low to medium-income groups in flood-prone areas who may loose stored food or assets; farmers who may have their land damaged or submerged by a rise in sea-level; and fishers who may loose their catch to shifted water currents or through flooded spawning areas. However, it is thought that by 2030 more countries will have improved their economies, infrastructure and institutions, and will be capable of compensating for the impact of climate change on domestic production through food imports from elsewhere.

[www.fao.org/DOCREP/MEETING/006/Y9151e.HTM]

AGRICULTURE'S ROLE IN MITIGATING CLIMATE CHANGE

Agriculture is itself responsible for about a third of greenhouse-gas emissions. Activities such as ploughing land and shifting ('slash and burn') cultivation for agricultural expansion release CO_2 into the air. Much of the 40 percent of human caused methane comes from the decomposition of organic matter in flooded rice paddies. About 25 percent of world methane emissions come from livestock. In addition, agriculture is responsible for 80 percent of the human-made nitrous-oxide emissions through breakdown of fertilizer and that of manure and urine from livestock. However, agriculture's GHG emissions can be largely reduced, and much can be done to lessen their effect on production and on the livelihoods of farmers, especially in developing countries.

Farmers can adopt coping mechanisms that withstand climate variability through activities such as the use of drought-resistant or salt-resistant crop varieties, the more efficient use of water resources, and improved pest management. Changes in cultivation patterns can include the reduction of fertilizer use, the better management of rice production, the improvement of livestock diets and the better management of their manure. In addition, national governments have an important role to play in enforcing land use policies which discourage slash and burn expansion and extensive (rather than intensive) livestock rearing, as well as raising the opportunities for rural employment.

Carbon sequestration can also be a means through which agriculture can make a positive contribution towards mitigation, and will be of growing economic and environmental importance in the context of the Kyoto Protocol. It is estimated that for the next 20 to 30 years, cropland contribution to carbon sequestration lies within the range of 450 – 610 million tonnes of carbon per year. By applying improved land management practices (better soil fertility and water management, erosion control, reversion of cropland in industrial countries to permanent managed forests, pastures or ecosystems, biomass cropping, conservation tillage, etc.), the role of agriculture as a major carbon sink and as a compensating mechanism for agriculture's contribution to GHGs can be greatly enhanced.

Agriculture can also play a role in reducing the burning of fossil fuels. Up to 20 percent of fossil fuel consumption could be replaced in the short term by using biomass fuel. In Brazil 6 million cars are running partly on alcohol derived from sugar cane. China already has 10 million dung digesters which provide a clean cooking fuel and an organic fertilizer. Fast-growing grasses, oilseeds and agricultural residues offer great potential as energy alternatives. It is important to note that these bioenergy initiatives also have a positive impact on rural socio-economic development.

Policy response can not only enhance agriculture's mitigating role, but at the same time it can reduce the vulnerability of poor people to food insecurity. New rural employment opportunities can be generated in efforts to replace fossil fuels with bioenergy. In addition, carbon sequestration programmes can help boost agricultural production as well as improve its overall sustainability. Regardless of the approach, technological and institutional changes must take place now before the impact of climate change becomes irreversible. But most importantly, poverty must be addressed and alleviated if the effects of climate change by the end of the next century are truly to be abated.

[www.fao.org/DOCREP/MEETING/006/Y9151e.HTM]

FAO ENVIRONMENT AND NATURAL RESOURCES SERIES

1. Africover: Specifications for geometry and cartography, 2000 (E)
2. Terrestrial Carbon Observation: The Ottawa assessment of requirements, status and next steps, 2002 (E)
3. Terrestrial Carbon Observation: The Rio de Janeiro recommendations for terrestrial and atmospheric measurements, 2002 (E)
4. Organic agriculture: Environment and food security, 2003 (E and S)
5. Terrestrial Carbon Observation: The Frascati report on in situ carbon data and information, 2002 (E)
6. The Clean Development Mechanism: Implications for energy and sustainable agriculture and rural development projects, 2003 (E)*
7. The application of a spatial regression model to the analysis and mapping of poverty, 2003 (E)
8. Land Cover Classification System (LCCS), version 2, 2005 (E)
9. Coastal Gtos. Strategic design and phase 1 implementation plan, 2005 (E)
10. Frost Protection: fundamentals, practice and economics- Volume I and II + CD, 2005 (E and S**)
11. Mapping biophysical factors that influence agricultural production and rural vulnerability, 2007 (E)
12. Rapid Agricultural Disaster Assessment Routine (RADAR), 2008 (E)
13. Disaster risk management systems analysis: A guide book, 2008 (E and S)
14. Community Based Adaptation in Action: A case study from Bangladesh, 2008 (E)
15. Coping with a changing climate: considerations for adaptation and mitigation in agriculture, 2009 (E)

Ar Arabic	**F** French	**Multil** Multilingual
C Chinese	**P** Portuguese	* Out of print
E English	**S** Spanish	** In preparation

The FAO Technical Papers
are available through the authorized
FAO Sales Agents or directly from:

Sales and Marketing Group - FAO
Viale delle Terme di Caracalla
00153 Rome - Italy

Printed on ecological paper